CAMBRIDGE LIBRARY COLLECTION

Books of enduring scholarly value

Physical Sciences

From ancient times, humans have tried to understand the workings of the world around them. The roots of modern physical science go back to the very earliest mechanical devices such as levers and rollers, the mixing of paints and dyes, and the importance of the heavenly bodies in early religious observance and navigation. The physical sciences as we know them today began to emerge as independent academic subjects during the early modern period, in the work of Newton and other 'natural philosophers', and numerous sub-disciplines developed during the centuries that followed. This part of the Cambridge Library Collection is devoted to landmark publications in this area which will be of interest to historians of science concerned with individual scientists, particular discoveries, and advances in scientific method, or with the establishment and development of scientific institutions around the world.

The Works of John Playfair

John Playfair (1748–1819) was a Scottish mathematician and geologist best known for his defence of James Hutton's geological theories. He attended the University of St Andrews, completing his theological studies in 1770. In 1785 he was appointed joint Professor of Mathematics at the University of Edinburgh, and in 1805 he was elected Professor of Natural Philosophy. A Fellow of the Royal Society, he was acquainted with continental scientific developments, and was a prolific writer of scientific articles in the *Transactions of the Royal Society of Edinburgh* and the *Edinburgh Review*. This four-volume edition of his works was published in 1822. Volume 1 includes a biography of Playfair, and his *Illustrations of the Huttonian Theory of the Earth*, which did much to popularise Hutton's ideas.

Cambridge University Press has long been a pioneer in the reissuing of out-of-print titles from its own backlist, producing digital reprints of books that are still sought after by scholars and students but could not be reprinted economically using traditional technology. The Cambridge Library Collection extends this activity to a wider range of books which are still of importance to researchers and professionals, either for the source material they contain, or as landmarks in the history of their academic discipline.

Drawing from the world-renowned collections in the Cambridge University Library, and guided by the advice of experts in each subject area, Cambridge University Press is using state-of-the-art scanning machines in its own Printing House to capture the content of each book selected for inclusion. The files are processed to give a consistently clear, crisp image, and the books finished to the high quality standard for which the Press is recognised around the world. The latest print-on-demand technology ensures that the books will remain available indefinitely, and that orders for single or multiple copies can quickly be supplied.

The Cambridge Library Collection will bring back to life books of enduring scholarly value (including out-of-copyright works originally issued by other publishers) across a wide range of disciplines in the humanities and social sciences and in science and technology.

The Works of John Playfair

VOLUME 1

JOHN PLAYFAIR
EDITED BY JAMES G. PLAYFAIR

CAMBRIDGE UNIVERSITY PRESS

Cambridge, New York, Melbourne, Madrid, Cape Town,
Singapore, São Paolo, Delhi, Tokyo, Mexico City

Published in the United States of America by Cambridge University Press, New York

www.cambridge.org
Information on this title: www.cambridge.org/9781108029384

© in this compilation Cambridge University Press 2011

This edition first published 1822
This digitally printed version 2011

ISBN 978-1-108-02938-4 Paperback

THE

WORKS

OF

JOHN PLAYFAIR, ESQ.

&c. &c. &c.

Printed by George Ramsay and Company.

THE

WORKS

OF

JOHN PLAYFAIR, ESQ.

LATE

PROFESSOR OF NATURAL PHILOSOPHY IN THE UNIVERSITY OF
EDINBURGH,

PRESIDENT OF THE ASTRONOMICAL INSTITUTION OF EDINBURGH,

FELLOW OF THE ROYAL SOCIETY OF LONDON,

SECRETARY OF THE ROYAL SOCIETY OF EDINBURGH,

AND HONORARY MEMBER OF THE ROYAL MEDICAL SOCIETY OF
EDINBURGH.

WITH

A MEMOIR OF THE AUTHOR

VOL. I.

EDINBURGH:

PRINTED FOR ARCHIBALD CONSTABLE & CO. EDINBURGH,

AND HURST, ROBINSON, AND CO. LONDON.

1822.

ADVERTISEMENT.

In the following Volumes are contained all the publications to which Mr PLAY-FAIR affixed his name, with the exception of the Elements of Geometry, and of the Outlines of Natural Philosophy, which were intended only for the use of Students, and although excellently adapted to their object, would possess but little interest for the general reader.

To Mr PLAYFAIR's acknowledged works has been added a selection from his contributions to the Edinburgh Review; those articles being chosen which contain a discussion of the higher parts of Physical Science, rather than of the merits of any individual Author.

Some few papers which were too valuable to be suppressed, although not so highly finished as to be published separately, have been united to the Biographical Memoir, which is prefixed to this Volume. The admirable delineation of Mr PLAYFAIR's Character, which forms the conclusion of that Memoir, is from the pen of FRANCIS JEFFREY, Esq.; and in the Appendix will be found a Letter on the same subject by Mr DUGALD STEWART. For the remainder of the Memoir, and for the execution of the whole Work, the Editor is alone responsible.

JAMES G. PLAYFAIR.

EDINBURGH, *Jan.* 8, 1822.

CONTENTS

OF

VOLUME FIRST.

———

BIOGRAPHICAL ACCOUNT

OF THE LATE

PROFESSOR PLAYFAIR.

BIOGRAPHICAL ACCOUNT

PROFESSOR PLAYFAIR.

JOHN PLAYFAIR was the eldest son of the Reverend James Playfair, minister of Benvie in Forfarshire, and was born at that place on the 10th March 1748.

He was educated at home by his father till he reached the age of fourteen, when he was sent to the University of St Andrew's, to prosecute his general studies, and to qualify himself for the church ; the profession for which he was intended. . Here his genius and uncommon application to study soon attracted the notice, and gained him the friendship, of his instructors. So remarkable, indeed, was his progress in the mathematical sciences, that Professor Wilkie, when confined by illness, selected him as the person best qualified to deliver the Lectures

on Natural Philosophy ; and, notwithstanding the
great disparity of years between the Professor and
the student, they became intimate friends.

In the year 1766, he distinguished himself in a
still more public manner, as a candidate for the Pro-
fessorship of Mathematics in the Marischal College
of Aberdeen ; and, although only eighteen, sustained with the greatest credit a trial which lasted eleven
days. Of six candidates, two only stood before him,
the Reverend Dr Traill, who was appointed to the
chair, and Dr Hamilton, who now fills it. The following extract from the conditions presented to the
candidates before the trial, will show the extent of
mathematical knowledge requisite to afford any hope
of success.

" Each of the candidates is to demonstrate some
of the propositions in each of the first six books of
Euclid, and any of the first twenty-two propositions
of the eleventh book. The candidates are to demonstrate propositions in plane and spherical trigonometry, and to apply the propositions to the actual solution of cases, and to explain the orthographic, stere-
ographic, and gnomonic projections of the sphere.
They are further to explain the genesis of the three
conic sections, and to demonstrate their capital properties. The candidates are to have questions put
to them relating to the principles of algebra, the nature and composition of equations, and their resolution by the method of divisors, and other methods ;

the arithmetic of surds, the composition of powers, and extraction of roots, the doctrine of ratios, the method of exhaustion as used by the ancients, the method of indivisibles, the arithmetic of infinites, the doctrine of prime and ultimate ratios, and the method of fluxions, direct and inverse, the nature of logarithms, and the expression of fluents by the measures of ratios and angles." *

It must be allowed that no ordinary union of industry, and of talent, was requisite to attain such an extensive knowledge of mathematics at so early a period of life.

In 1769, having finished his studies, he quitted the University, and for some years spent much of his time in Edinburgh, chiefly in the society of Dr Robertson the historian, Adam Smith, Dr Matthew Stewart, Dr Black, and Dr Hutton.

It would appear from letters published in the Life of the late Principal Hill, that, during this time, Mr Playfair had twice hopes of obtaining a permanent situation. The nature of the first, which offered itself in 1769, is not there specified, and is not known

---

* The particular questions proposed to the candidates for solution, were such as to require a complete command of each of the subjects above enumerated.

We have to acknowledge the kindness of Dr Hamilton and Dr Brown of Aberdeen, for an account of the trial, as preserved in the records of the University.

to any of his own family; the second was the Pro-
fessorship of Natural Philosophy in the University
of St Andrew's, vacant by the death of his friend
Dr Wilkie, which took place in 1772. In this,
which he earnestly desired, and for which he was
eminently qualified, he was disappointed; "the situation," to use the words of Dr Cook, "being con-
ferred upon another gentleman, one of their own
number, who had so powerful a claim upon them,
that Lord Kinnoul mentions to Mr Hill, that, had
Mr Playfair known of the wish of this gentleman to
succeed Dr Wilkie, he would not have become a
candidate."

In the course of the same year this object was
rendered still more desirable, by the death of his
father, an event which devolved upon him the charge
of his mother and family, of whom one brother only
was sufficiently advanced to be independent. Nearly a year, however, elapsed before his wishes were
accomplished; for although Lord Gray immediately
presented him to his father's livings of Liff and
Benvie, yet that nobleman's right of presentation
was, in this instance, disputed by the Crown Law-
yers: and it was not till August 1773 that he obtained possession by a resolution of the General As-
sembly of the Church, for which he was chiefly indebted to the strenuous support of his friend Dr
Robertson. The legal question continued long
dependent before the Court of Session, but was

finally decided in favour of Lord Gray, by which his nomination was confirmed.

Mr Playfair now became resident at Liff, where he devoted the chief part of his time to the duties of his charge, composing for it many sermons in the simple and convincing style of eloquence by which his writings are so strongly characterized; while his leisure hours were filled up with the superintendence of the education of his brothers, and the prosecution of his own studies. His correspondence of this date, with his friend Mr Robertson, (now Lord Robertson,) shows a most remarkable extent of reading, and contains a discussion of the merits and opinions of Macchiavelli, Locke, Leibnitz, Helvetius, Reid, Sextus Empiricus, Plato, Bacon, Price, Cudworth, Boscovich, Priestley, Johnson, Beattie, and Hartley; an account and refutation of the attempt to explain gravitation by an ethereal fluid, and many ingenious observations upon the geography and the singular social institutions of the South Sea Islands, then recently discovered. *

Beside occasional visits to Edinburgh, he made an excursion in 1774 to Perthshire, where Dr Maskelyne was then engaged in a set of experiments on the effect of mountains in disturbing the direction of the plumb-line; and during a short stay on the

---

* For the perusal of these very interesting letters, and for permission to make use of the information contained in them, we are indebted to the kindness of Lord Robertson.

side of Schehallien, a friendship was formed which terminated only with the life of the Astronomer Royal. Under his auspices, four years later, Mr Playfair's first mathematical essay, on the Arithmetic of Impossible Quantities, was presented to the Royal Society of London, and published in 1779, in the sixty-eighth volume of the Philosophical Transactions.

In 1782 he was induced, by very advantageous offers, to resign his living for the purpose of superintending the education of the present Mr Ferguson of Raith and his brother, Sir Ronald Ferguson. He was thus enabled to reside, at least during the winter, in Edinburgh, and to enjoy the literary society in which he was himself so well qualified to shine. He also profited by an interval of leisure to pay his first visit to London, where, through the kindness of Dr Maskelyne, he found a ready introduction to the scientific world. *

In 1785 an exchange took place between Dr Adam Ferguson, Professor of Moral Philosophy in the University of Edinburgh, and Mr Dugald Stewart, who then filled the Mathematical chair; and the former, being disabled by delicate health, from going through the laborious duties of his office,

---

* Of this visit he drew up, on his return to Scotland, a minute account, extracts of which will be found in the Appendix, No. I.

Mr Playfair was appointed Joint Professor of Mathematics, a situation in which he remained for twenty years. He, nevertheless, continued with his two pupils until the year 1787, when his charge having terminated, he joined his mother and sisters, who had for some years been resident in Edinburgh.

He now wrote and published successively, in the Transactions of the Royal Society of Edinburgh, a Life of Dr Matthew Stewart, the late Professor of Mathematics in the University of Edinburgh ; a paper on the Causes which affect the Accuracy of Barometrical Measurements; Remarks on the Astronomy of the Brahmins; and a paper on the Origin and Investigation of Porisms ; of which a second part, containing their algebraic analysis, was promised, but never appeared, nor are there any traces of his having prosecuted the subject any farther.

His studies, indeed, at this period (1793) experienced a partial interruption from the sudden death of his brother James, established in London as an architect. Mr Playfair did not hesitate, upon this occasion, to undertake an office the most foreign to his habits of life, namely, that of arranging complicated business ; and besides showing the greatest kindness to the whole family, he, in the following year, adopted and took under his own care the eldest son, then only six years of age.

In 1795, he published an octavo volume, containing the Elements of Geometry as he taught them in his class. The chief peculiarity of it is, the introduction of algebraic signs in the fifth book, in order to render the proportions more compact, and consequently more easily followed by the eye. To the first six books of Euclid, thus modified, were added by Mr Playfair three containing the rectification and quadrature of the circle, the intersection of planes, and the geometry of solids; then follow Plane and Spherical Trigonometry, with the Arithmetic of Sines. The notes, which are peculiarly valuable, are thrown into the form of an appendix, and contain the author's reasons for the alterations made in the various parts of the volume, and a discussion of the difficult subject of parallel lines. It is worthy of notice, as a proof of the high estimation in which this work is generally held, that it has gone through five editions of a thousand copies each; and that four of these editions were called for while it was not taught in the University of Edinburgh.

During the winter of 1797, while confined by a severe attack of rheumatism, he amused himself by keeping a journal of his studies, from which it appears, that Physical Geography and Climate, the law which regulates the decrease of temperature observed in ascending into the higher regions of the atmosphere, and the influence of that decrease on barometrical measurements, were the favourite objects of his attention. His less serious pursuits consisted in reading voyages and travels, in sketching an analytical treatise on the Conic Sections, and in composing an Essay on the accidental discoveries which have been made by men of science, whilst in pursuit of something else, or when they had no determinate object in view. At this time, also, were written the Observations on the Trigonometrical Tables of the Brahmins, and the Theorems relating to the Figure of the Earth, afterwards published in the Transactions of the Royal Society of Edinburgh.

In the spring of the same year, however, a new direction was given to his thoughts, by the death of his esteemed friend Dr James Hutton, of whose works he began to draw up an abstract, with a view to the composition of a biographical memoir; an occupation which eventually gave birth to the Illustrations of the Huttonian Theory of the Earth.

Two powerful reasons concurred to induce Mr Playfair to present his friend's theory to the world in a form different from the original. The peculiar style of composition and arrangement adopted by Dr Hutton, both in the sketch published in the Transactions of the Royal Society of Edinburgh, and in the more extended work which followed, rendered his theory less intelligible, and much less known than its merits deserved. The same cause gave rise to many misrepresentations and attacks from the few who had read it. To afford a clear exposition of the theory, and to repel these attacks, were the objects of the Illustrations, and from which that work derived the form in which it appeared; namely, a series of chapters stating the positions of the Huttonian theory, the facts which make for it, and the arguments which have been urged against it. With what success this was attended we may judge from the fame and credit which have been attained by the theory, which, but for its commentary, seemed likely to be known only through the erroneous statements of its opponents.

We have often heard the Illustrations quoted as a model of purity of diction, simplicity of style, and clearness of explanation; but with a regret, that such powers were employed on a subject so unsatisfactory as a theory of the earth. But the Huttonian theory, contented to explain the changes which take place in the crust of the globe, is wholly free from the reproach so justly attached to those which extend to the original creation of the world; and it is remarkable as the only one which agrees with physical astronomy, in assigning to our system a principle of compensation which leaves its duration limited only by the will of its Creator. If it is intended to censure as unphilosophical every attempt to explain the structure of the earth, we should suggest that some respect was due to a study which found favour with such men as Hutton, Black, and Playfair.

No less than five years, from 1797 to 1802, had been occupied in writing the Illustrations, and it was not till 1803, that, in the Transactions of the Royal Society of Edinburgh appeared the Biographical Sketch, in which Mr Playfair did ample justice to the powerful and comprehensive talents of his friend. It was, indeed, a subject calculated to call forth his highest powers; his kindness of heart, and his admiration of genius, both conspired to animate his pen, of which it is perhaps one of the finest productions.

In 1805, he quitted the Mathematical chair, to succeed Professor John Robison in that of Natural Philosophy. As Professor of Mathematics he had exerted himself to the utmost to inspire his students with a taste for the science; and as the most effectual mode of doing so, and at the same time of ascertaining their progress while under his own care, he proposed numerous exercises for solution, and rewarded the diligent by naming them before the class. In this way he encouraged habits of investigation of the greatest value, even to those who regarded mathematics only as a part of their general education; while, for the benefit of those who wished to cultivate the higher branches of the science, he taught at intervals a third class, rendered doubly valuable by his intimate and masterly knowledge of the modern analysis, at that time so little attended to in Britain. This class was attended by many who had long finished their academical studies, and who testified their sense of obligation to their instructor by presenting him with a valuable astronomical circle, now placed in the observatory of the Astronomical Institution.

The extensive subject of natural philosophy upon which he was now to enter, afforded him an opportunity of displaying more fully those qualities which rendered his lessons at once so instructive and so attractive. His lectures on the appearances of the planetary system were distinguished for their eloquence, while the most abstruse propositions of physical astronomy and of optics were established by demonstrations of a simple and elementary nature. He possessed, indeed, in a degree rarely to be met with, the art of facilitating to others the attainment of that knowledge which he had himself acquired by profound study.

His appointment having taken place in spring, he retired during the summer to Burntisland, a village in the neighbourhood of Edinburgh, that he might devote his whole attention to the preparation of his lectures. His mother had died in the preceding year, and his eldest sister having quitted him, his family now consisted of his youngest sister and two nephews; one already mentioned, and a younger brother, whom he had also taken under his own care.

The well known disputes which took place concerning the appointment of a successor to Mr Playfair, induced him to address the Lord Provost, as

chief patron of the University, in an animated letter,
vindicating the rights of science ; in consequence of
which, in the controversy that ensued, an attack was
made upon himself, which could not be passed over
in silence, and which he repelled, in an answer remarkable for keenness of retort and force of reasoning.   This unpleasant interruption being, however,
soon at an end, he returned to studies more con-
genial to his feelings ;   the investigations which
gave rise to the essays on the Solids of greatest attraction, and on the Progress of heat in spherical
bodies, which appeared, at a later period, in the
Transactions of the Royal Society of Edinburgh.
During this time, also, he presented to the Royal
Society of London (of which he had been elected
a Fellow in 1807) the account of the Lithological
Survey of Schehallien.

In 1814, he published, for the use of his students,
Outlines of Natural Philosophy, in two volumes octavo.   The first volume treats of Dynamics, Mechanics, Hydrostatics, Hydraulics, Aerostatics, and
Pneumatics ;   the second is entirely devoted to Astronomy.   Optics, Electricity, and Magnetism, were
to be comprised in a third volume, which would have
completed the work, but was never executed.   Being
intended to present merely an outline of natural
philosophy, the propositions are in general given
without demonstrations, but with a reference to the
more extended works in which these are to be found ;

and, in every case, is subjoined the formula by which the result may be applied to practice. In the following year appeared, in the Transactions of the Royal Society of Edinburgh, the Life of Professor Robison, who had been his predecessor in the office of Secretary to that institution, of which Mr Playfair was a most active and zealous supporter. Upon him chiefly devolved the task of arranging and publishing its Transactions, which he enriched with numerous papers, and a set of meteorological tables, from his own observations, since quoted by Humboldt, in the work on Isothermal Lines, as of the greatest value.

Besides the various publications just mentioned, and numerous contributions to the Edinburgh Review, two works of great importance had for some years occupied Mr Playfair's most serious attention. One of these was the Dissertation on the Progress of Mathematical and Physical Science since the Revival of Letters in Europe, written for the Supplement to the Encyclopædia Britannica ; and published in that work in 1816. The other, which was the first conceived, but interrupted in the execution by the Dissertation, was a second edition of the Illustrations of the Huttonian Theory of the Earth. This edition, of much greater magnitude than the former, was likewise completely different in the arrangement of its contents. It was intended to commence with a description of all the well authenticated facts in

1

geology collected during his extensive reading and
personal observation, without any mixture of hypothesis whatever.   To this followed the general in-
ferences which may be deduced from the facts, an
examination of the various geological systems hitherto offered to the world, and the exclusion of
those which involved any contradiction of the principles previously ascertained; while the conclusion
would have presented the developement of the system adopted by the author, and the application of it
to explain the phenomena of geology.   It must be
viewed by every one as a great loss to science that
this design was never completed, for such an analysis
of voyages and travels, such a description of geological phenomena, such a system of physical geography,
as would have been contained in the first division of
this work, we can scarcely hope to see ; and where
is to be found the geologist who will bring to the
execution of the theoretical part, the candour in the
search of truth, the habits of accurate reasoning, and
the power of employing the mathematical sciences
as a test of the soundness of his conclusions, all possessed in so high a degree by Mr Playfair ?   Qualities which have stamped his geological speculations
with this peculiar character, that, whether we are
disposed entirely to adopt them or not, we are certain that they contain nothing inconsistent with the
laws of the physical world : a character by which,

VOL. I.   c

perhaps, more than any other, they are distinguished from those of geologists in general.

The prosecution of his geological studies led Mr
Playfair to spend a portion of the summer vacation,
almost every year, in travelling through the more interesting districts of England, Scotland, and Wales,
of which he thus acquired a most accurate knowledge. These researches he had long been desirous
of extending to the Continent, and had, in 1802,
nearly completed arrangements to that effect, when
the sudden renewal, and long continuance of the
war, destroyed, to all appearance, every prospect of
accomplishing his wishes. He then turned his attention to Ireland, and had visited Dublin and the
Giant's Causeway, when the general peace of 1815
enabled him to resume his former plan of an extensive journey through France and Switzerland; to be
prolonged, if he could obtain leave of absence for a
winter, to the southern extremity of Italy. To examine the geological structure of such an extent of
country was no small undertaking at the age of six-
ty-eight; but by regular exercise and frequent excursions, Mr Playfair had preserved a degree of activity and a power of exertion rarely to be met with
in literary men at a much earlier period of life, while
he was relieved from the details of the journey by
his eldest nephew, then just returned from a resid
ence of some years in the Mediterranean.

The following account of the journey is given at considerable length as the only memorial of observations from which so much was to be expected, and is drawn up from Mr Playfair's notes, made in pencil on the spot. The greater part of these, as he seldom wrote down the reasonings suggested by the facts which he observed, however interesting they might have been when incorporated with the second edition of the Huttonian Theory, for which they were intended, would here present a dry detail of the names of rocks, and the position of strata; but wherever the information they contain is either new, or of superior accuracy, it has been extracted, and, if necessary, given in his own words. The conversations and intercourse which passed between him and the many individuals of distinction with whom he became acquainted, are passed over in silence, as, however interesting they might prove, they could not be made public without a breach of confidence; which, although not unfrequently committed by travellers, is both reprehensible in itself, and injurious to society.

Of his residence during six weeks at Paris, we have, therefore, little to remark, except the ease with which, after every introduction, the transition was made from the first sentences of ceremonious welcome, to the free exchange of opinions on every subject. To this the language presented no obstacle; extensive reading enabled Mr Playfair to express himself freely in French on subjects of science, and it was only in the lighter conversation, and in the common occurrences of life, that he required any assistance. Besides the enjoyment of society, and the objects of general interest so numerous in that metropolis, he found an opportunity of studying the minerals and rocks of France in the museum of the Ecole des Mines; and had the pleasure of examining the basin of Paris in company with Cuvier and Brongniart, the porphyries and volcanic productions of the Andes with Humboldt and Bonpland. It would, indeed, be a most culpable omission, were we not to record the strong sense entertained by Mr Playfair of the flattering reception and kind attention he experienced from the literary and scientific society of Paris.

From Paris he proceeded by Fontainebleau, Dijon, Besancon, and Pontarlier, to Neuchâtel in Switzerland. This line of road presents nothing remarkable, and it was only on entering the Val Travers in Mont Jura, that he met with a phenomenon, curious in itself, and which had often engaged his attention, namely, the existence of loose blocks of granite, gneiss, and mica slate, on the surface of a chain of mountains entirely calcareous. They first appeared within the French frontier on the western declivity of Jura, and numbers are scattered all the way through the defiles to Neuchâtel. The largest and most striking of them, the Pierre Abot, (so named from the farm in which it is situated,) lies concealed in a wood upon the rapid slope of a hill, at an elevation of at least 700 feet above that town, and measures 64 feet in length, 32 in breadth, and 16 in height. It contains, therefore, 32768 cubic feet, which (allowing 13 cubic feet of granite to a ton) gives a weight of 2520 tons.

When we consider that the nearest point where the granite is to be found in its native place, is at a distance of 70 miles, it will appear no easy matter to assign a conveyance by which this block could have performed such a journey over intervening hills and vallies without considerable injury. A current of water, however powerful, could never have carried it up an acclivity, but would have deposited it in the first valley it came to, and would in a much less distance have rounded its angles, and given to it the shape so characteristic of stones subjected to the action of water. A glacier, which fills up vallies in its course, and which conveys the rocks on its surface free from attrition, is the only agent we now see capable of transporting them to such a distance, without destroying that sharpness of the angles so distinctive of these masses. That mountains formerly existed of magnitude sufficient to give origin to such extensive glaciers, is countenanced by other phenomena observed in the Alps, and does not imply any alteration in the surface so great as the supposition of a continued declivity between the two extreme points, which is, after all, insufficient to remove the objection arising from the sharp angles of these rocks.

From Neuchâtel, Mr Playfair prosecuted his journey through the low country of Switzerland by Bienne, Soleure, Arau, and Baden, to Schaffhausen, at its northern extremity, and thence proceeded to Geneva, passing through Zurich, Lucerne, Berne, Morat, and Lausanne. From Lucerne he intended to make an excursion among the mountains, but the unceasing rains rendered the attempt hopeless, and he was compelled to give it up, after a delay of eight days. The same cause obliged him to forego one of the great objects of a visit which he made to the valley of Chamouni, soon after his arrival at Geneva. After enjoying two fine days while examining the valley and its environs, he set out to visit the rocks which have been the subject of so much discussion under the name of the puddingstone of Valorsine ; but a thunder-storm, with heavy rain and dense mist, came on while he was on a high plain to the left of the Col de Balme, and it was not without much difficulty, after a search of more than an hour, that the guide regained the path which led back to the valley of Chamouni. This being accomplished, he remained all night in a cottage at the village of La Tour, and proceeded next morning by the Tête Noire to Trient, and thence to Martigny in the Valais. At Trient he had the satisfaction of finding several fallen masses of the rocks which he had

wished to examine, and which are situated above
Trient at a height of nearly 3000 feet. The following are his own notes made at the time : " They
are decided puddingstones ; the included stones are
rounded and water worn, consisting chiefly of feldspath, quartz, and petrosilex. Nobody can possibly
mistake them for any thing but an agglomerate of
the kind just mentioned. I saw none of them in
their place, but one block was so large as to show
that the stratification was vertical. The cementing
substance is a mica slate, and it appears that the
planes of the mica are parallel to those of the beds
of the stone."

Although he did not see the rock in its native
bed, there could not be any mistake as to its
identity ; for his guide was old Balma, the com-
panion of Saussure, and who pointed out very
precisely, in this excursion, specimens of various
minerals, with the names given to them by that author. Mr Playfair and Balma derived much pleasure from the society of one another, and parted
with mutual regret. The works of Saussure had
been studied by Mr Playfair for years, and he was,
as it were, already well acquainted with Balma, who,
on the other hand, was delighted at meeting with
one who possessed that knowledge which he esteemed most highly, and who was animated by a similar
admiration for his late master. It is but justice to
the family of M. de Saussure to add, that Balma,

although he still follows the occupation of a guide, is considered by them as under their protection, from which he can at any time derive whatever assistance he may require. From Martigny Mr Playfair returned to Geneva, by St Maurice and the eastern side of the lake.

The season was now far advanced, and he had the prospect of returning home without having been able to visit any of the central portion of the Alps, when he received a letter from the Lord Provost of Edinburgh, granting him leave of absence for the ensuing winter : this, in fact, left him at liberty for a year, enabling him to accomplish his wish of passing the winter in Italy, and of revisiting Switzerland and France in the following summer. His academical duties had been readily undertaken by his colleague, Professor Leslie.

After a residence of nearly a month, during which he had renewed many agreeable acquaintances formed in Edinburgh, Mr Playfair quitted Geneva to enter Italy by the Simplon. The road, which, as far as Martigny, he had already seen on his return from Chamouni, continues throughout in the Valais, where, as in all the longitudinal vallies of a country constructed on such a vast scale, the strata present little or no variety. Much of the ascent of the Simplon, from the Swiss side, lies upon a tender schistus, subject to great slips, equally injurious whether they take place above or below the road, and very difficult to con-

trol. Upon the schistus rests a strong gneiss, of
which the beds are often broken in pieces by their
own weight, when deserted by the feeble rock below. On the south side the road sinks rapidly down
to the bed of the Vedro, which it follows through a
succession of wild and singular defiles, between perpendicular cliffs of great height, where all the accidents
to which hard rocks are subject are fully exemplified.
On the Swiss side, the excellence of the line of road
is entirely due to the skill of the engineers who
traced it ; on the Italian side its course was pointed
out by the river, but the difficulties to be overcome,
and the works to be executed, were of much greater
magnitude ; the galleries cut through the solid rock,
(gneiss,) three in number, being, together, not less
than 900 feet in length, every where 30 feet in
height, and 25 feet in width.

At Baveno, on the Lago Maggiore, are extensive
quarries of a beautiful pale red granite, much employed in the public buildings at Milan, and which
have often been cited as affording an instance of
the stratification of that rock. The external surface
is undoubtedly marked with the appearance of seams,
but no one of them can be traced to any extent,
and where the rock is cut into by the operations of
the quarry, they disappear altogether. This is a
complete proof that they are accidental marks, produced by external causes, as it is well known that

true stratification is more distinct in the interior than at the surface.

At Milan, he found ample occupation in the collections of Father Pini, of Breislac, and of the Council of Mines, rich in the fossil bones of Lombardy, the shells of the Apennines, and the volcanic productions of Naples, while his evenings were enlivened by the society of Breislac, Brocchi, De Cesaris, and Oriani. He made, also, an excursion to the upper extremity of the Lake of Como, in hopes of seeing the primary rock, but without success.

From Milan he proceeded by Pavia, Lodi, Piacenza, Parma, Reggio, Modena, and Bologna, to Florence. At Bologna, after visiting the meridian in the church of San Petronio, his attention was chiefly directed to the Institute, which possesses a fine apparatus for experimental philosophy, and a remarkable collection of anatomical preparations, executed in wax. Relieved from the disgust attendant upon the reality, he spent much time in studying the anatomy of the eye and ear, and expressed a strong desire to possess such preparations, for the use of his own class. The only person to whom he had an introduction, perhaps the most remarkable in Italy, was the librarian of the Institute, the Abate Mezzofanti, who is said to be master of more than thirty languages. His conversation showed him to be a man of great general knowledge, and the English, in which it was carried on, was excellent. As might be expected, it was more studied, and less familiar than that of an Englishman, but what was contrary to all expectation, was the great accuracy of the pronunciation.

At Florence, Mr Playfair made a stay of nearly three weeks, which were chiefly occupied in visiting the galleries and the museum ; but of all the numerous objects of admiration contained in them, none addressed themselves so forcibly to his feelings as the relics of the Academia del Cimento.   The room in which these philosophers met, the table round which they sat, and the instruments employed in their experiments, are all carefully preserved in their original condition.   Among the latter is the telescope of Galileo, made of two semi-cylinders of wood, four feet long, coarsely hollowed out, tied together with threads, and covered with paper.   He had, also, the pleasure of becoming acquainted with Sismondi, the distinguished historian of the republics of Italy.

On the 12th of November, he set out from Florence for Rome, following the eastern road, which passes through Arezzo, Perugia, Foligno, Spoleto, Terni, and Cività Castellana.

The limestone rock between Spoleto and Terni presents a most remarkable instance of inflected strata.   A number of contiguous beds rise from under the soil, truly vertical, and bend suddenly over at right angles into the horizontal position, without any curvature, any fracture, or any disturbance to the neighbouring strata.

At Terni appears for the first time the peculiar stalactitic rock, known by the name of Travertine, of which the principal formation is in the vicinity of Rome. It is here deposited by the Velino, which, descending from the elevated plain of Rieti into the valley of the Nera, five hundred feet deep, forms the fall of Terni, pre-eminent in height, the quantity of water, and the beauty of the surrounding scenery.

But it was after crossing the Tiber at the Ponte Felice, that Mr Playfair entered upon a country totally different from any he had ever seen. To use his own words : " As soon as we had reached the south side of the bridge, I was surprised to see a thick bed of strong stone lying along the top of a mass of loose gravel. This extraordinary appearance was in some degree resolved by the information which the hammer afforded, that this rock is in fact lava." A few miles farther on, close to Città Castellana, commences the volcanic tufa, which forms the basis of the country, and is here cut by a small stream to the depth of a hundred feet. Nothing can be more striking than these ravines, level at the top with the general surface of the country, and cut to so great a depth, their sides broken into slender pyramidal columns, and the banks of the stream

4

which runs in the bottom, ornamented with the evergreen oak and the arbutus.

At Rome, where he arrived on the 18th of November, Mr Playfair remained during the winter, and found in the remains of antiquity, the treasures of the Vatican, and the singular nature of the adjacent country, a combination of all that could afford him occupation or pleasure. When to this we add the enjoyment of a circle of English society, such as can rarely be formed out of London, and which comprised several of his own friends, we shall not be surprised that he always counted the winter at Rome among the happiest days of his life. The chief objects of geological research were the quarries of Capo di Bove, and the shells of Monte Mario. In the former he found a rock, which, from its even fracture, comparative lightness and porosity, and the variety and abundance of crystals observed in it, bore a greater resemblance to lava than to greenstone; but which, at the same time, contained much carbonate of lime in veins and in cavities, which had evidently no connection with the surface. The shells of Monte Mario are found at a height of from seven to eight hundred feet above the level of the sea, but in unconsolidated earth, and in their natural state; in many even the internal pearly coat remaining uninjured.

In these investigations he experienced the greatest kindness and attention from Professors Morichini

and Carpi; as also from Mr Niebuhr, the Prussian
Envoy at Rome, in his search for manuscripts in the
Vatican Library; which, however, proved fruit-
less, the MSS. of Diophantus being such as were already known to him, and that of Pappus Alexandri-
nus being no longer to be found.

From Rome he proceeded by the usual route to
Naples, where, in the study of an undoubtedly volcanic country, he was to acquire the knowledge
which might enable him to decide in doubtful cases.
Vesuvius itself, at that time active, and pouring out
considerable streams of lava, was the first object of
his attention. He then examined Monte Somma,
the currents of lava which are the produce of more
ancient eruptions, the Zolfatara, Monte Nuovo,
the numerous craters which surround the city of
Naples, and lastly spent three days in the island of
Ischia.

It is not from the cone of Vesuvius, which is in a
state of activity, and which produces the most forcible impression upon the imagination, that the most
valuable information is derived as to the mode in
which a volcano conducts its operations. The principal facts to be noticed with regard to it, are, that
the light proceeds not from flame, but from the reflection of the red-hot mass in the crater, that what
appears to be smoke is a cloud of fine cinders and
ashes, and that there is a discharge of elastic fluid
with each ejection of scoriæ. To attain a know-

ledge of the means by which a volcano elevates itself above the plain, and of its internal structure,
we must consult Monte Somma, upon which the
whole is impressed in very distinct characters.

Monte Somma is separated by a wide valley called the Atrio del Cavallo, from the cone of Vesuvius, which it surrounds from north-west to east.
The external surface of Somma, which is to the north,
has a slope like that of Vesuvius, is covered with se-
mivitrified, light, porous, black, volcanic cinders, and
is marked with numerous semi-cylindric undulations
running from top to bottom.   The internal surface
fronting Vesuvius is a perpendicular wall, many
hundred feet in height at the north-western extremity : the ridge, which is quite sharp, and affords a
very precarious and difficult path, sinks towards the
east, and at last vanishes altogether in the plain of
the Atrio del Cavallo.   Mr Playfair walked along
this ridge as far as a very deep ravine, called the
Canale d'Arena, where he descended, and followed
the foot of the perpendicular wall through its whole
extent.   This perpendicular face presents horizontal
beds of compact lava, separated by layers of volcanic
breccia, the whole traversed by veins or dikes of
compact lava highly inclined and deviating very
little from the perpendicular.

The beds of compact lava approaching more or
less to horizontality, are in fact oblique sections of
beds dipping much to the north.   This is sufficient-

ly well shown even by the inequalities of the internal
face of Somma, but is completely proved on going along
the ridge, and viewing the outward declivity of the
mountain, when you see that these beds lie nearly
parallel to the surface. They are of very unequal
thickness, and it is seldom that any one extends far
in the horizontal direction. They are of a very compact lava, in which three varieties are chiefly to be
distinguished. The strongest has a blue ground, an
even fracture, or one slightly conchoidal, is studded
with large well formed leucites, and numerous crys
tals of pyroxene : the surface when weathered is a
dirty brown. The second variety differs from the
former, in being a little less compact, containing the
leucites in very small crystals with abundance of
pyroxene. The third contains no leucites, hardly
any pyroxene, has very little of the crystalline character, and resembles some of the most earthy of the
greenstones.

The stone that separates these beds, and that
constitutes by far the greatest part of the mountain,
is a volcanic breccia, of a reddish colour, consisting
of pieces of lava, firmly united, though clearly distinct. The hard pieces of the stone are of all the
three kinds just mentioned ; the intermediate substance is either tufa, or scorious fragments of lava.
The surfaces, both upper and under, of the beds
above described, and which are in contact with this
breccia, are generally scorious, as may be observed

in all lavas, both ancient and modern, which have actually flowed from Vesuvius.   The variety of the breccia is so great, that it would require a great deal of time to become fully acquainted with it.   The softer parts are apt to crumble down into sand.

The vertical dikes are compact, of a blue colour, and contain many crystals of leucite and pyroxene. They are very various in breadth, but each dike preserves its own pretty regular throughout, with the exception of a few which taper upwards: while some are only a few inches in width, others are several feet; but all extend several hundred feet upwards, and all have their faces distinctly defined. They intersect one another, and cut through the beds first described, but without producing any shifts.   In one place is a centre, from which proceed five of them, diverging at various angles : one of these is covered at both sides with very perfect glass of a dark colour.

It would appear that the mountain, when entire, has been penetrated from below by the vertical dikes, and that, at a later period, the volcanic force has fairly blown away the summit, leaving the interior exposed, as it is now seen, in the perpendicular face of Somma.   That the vertical dikes could not have been formed on the surface is quite evident, for they often extend several hundred feet in the vertical plane, with the faces well defined, while they are not above six inches in width.

At a short distance above Resina, among the vineyards, is to be seen a stream of lava, which is said to be that of the eruption of 1631. It has been quarried down to the lower surface, presenting a face thirty feet in height, of which six feet from the top and three or four from the bottom are scorious, while the central portion is exceedingly compact : the earth upon which it rests is red, and has the appearance of being scorched. This lava is full of pyroxene, but very few leucites are to be seen in it, and it has many large cavities, some even eighteen inches in diameter, lined with tubercles of the lava itself. It is very sonorous when struck, and the fracture is even.

Close to the sea-side at La Scala, south from Portici, is an ancient lava bearing no date, which is also quarried, and shows a face 45 feet high, and more than a quarter of a mile wide. It is like the last, scorious both above and below, while the centre is compact. Besides much pyroxene and some olivine, it is remarkable, as containing the green oxide of copper, and many filamentous crystals like those of the Capo di Bove near Rome. The colour is greyish blue, the fracture pulverulent, and the stone much less tough than any greenstone.

Still farther along the coast is the stream of lava which in 1794 destroyed the village of Torre del Greco, and is still exceedingly rugged and desolate,

although grass is beginning to appear in the hollows
of the surface.   The fresh fracture is very black
and porous, contains much pyroxene well crystallized,
but neither olivine nor leucites.   In the Fossa
Grande, a deep ravine which descends from the her-
mitage on Vesuvius towards the north-west, is contained the stream of lava of 1767, still very rugged,
with the current of 1810 above it, separated only
by earth and scoriæ.   The right side of the ravine
as you ascend, is formed by four streams of ancient
lava, lying one above another, in a similar way,
with earth interposed between them, and all of
them scorious at both surfaces.   The left side of the
ravine is formed by a face of tufa, about 300 feet
high, in which are found imbedded the stones resembling mica slate which contain the Sommite, Vesu-
vian, and other crystals peculiar to the old eruptions
of the mountain.   The Fossa di Faraone, another
very deep ravine in this vicinity, consists entirely of
tufa, in which are included many nodules of limestone, some of them highly crystallized, and resembling a white marble, others like the common limestone of the Appennines.

The stream of lava which descends from the Zolfatara to the sea-shore near Pozzuoli, is totally different in appearance from the rest: it is greyish white,
and contains numerous transparent crystals of feldspath, formed in rectangular parallelepipeds of greater length than breadth, and greater breadth than

thickness. It is, however, scorious on the under surface as well as the upper.

With the exception of the great current of Arso, which resembles the recent productions of Vesuvius, the lavas of Ischia are closely allied to that of the Zolfatara, differing only in being much decomposed, and exfoliating in large plates. This property, indeed, gives a very peculiar aspect to the surface of the whole island, nothing like a firm rock being visible from the sea-shore to the summit of the mountain of Epomeo in the centre, and the beds of the winter torrents being consequently numerous and deep. It is only in the small craters which lie near the northern shore that the lava appears more nearly in its original state, when it is exactly like that of the Zolfatara. Great quantities of pumice and of pitchstone in round nodules are found scattered on the surface near the town of Ischia.

One character common to all the streams of lava above described, is the scorious or vesicular state of the stone, both where in contact with the air, and with the ground over which it flowed; a fact, as far as Mr Playfair knew, not remarked by any author, and of which he immediately perceived the value, as pointing out wherever it occurs that the rock derives its origin from a volcano.

Of all the various effects of volcanic force so fully exemplified in the country round Naples, none is more impressive than the destruction of the ancient cities of Pompeia and Herculaneum. The former is covered to the depth of thirty feet with a mass of which the lowest stratum consists of pumice, and the upper of unconsolidated tufa, while the latter is imbedded in a hard solid tufa rock of twice that depth, above which lies a stream of compact lava. The former has evidently been destroyed by a shower of volcanic ashes; but it is difficult to conceive how the latter was enveloped by a substance so soft and plastic as to receive an impression, so hot as to char vegetable matter, and yet capable of forming a rock, which, although sufficiently porous to allow the percolation of water, is so hard as to render its removal a work of great difficulty.

Having thus accurately examined the volcanic phenomena of the Neapolitan territory, Mr Playfair, after an excursion to Pæstum, set out on his return to Rome. In crossing the insulated group of hills, of which Monte Cavo is the centre, he observed near Albano a rock which, although closely resembling greenstone even in the fracture, and containing calcareous matter, proved, upon a more close examination, to be a lava reposing upon alluvial earth.

At Rome he remained only long enough to visit Tivoli, and to investigate the singular formation of travertine, which stretches from that place over a great portion of the plain. Travertine, called by the ancients Marmor Tiburtinum, is a calcareous

stone, of which the deposition is constantly going on
at the surface, but very different in every respect
from an ordinary stalactite. It is of a warm cream
colour, and so full of air-holes as to deserve the
name of cellular, yet so hard and imperishable as
to have been employed in building St Peter's and
the Colosseum. The quarries from which the materials of these great edifices were obtained, lie in
the plain three miles and a half from Tivoli, and
about fourteen from Rome. They are very extensive, but quite superficial, never exceeding a depth
of fifteen feet; a limit which even now is never passed, for want of a contrivance to get rid of the water
which every where springs up. The rock lies in horizontal beds, from which blocks of almost any magnitude can be raised by wedges, and as it recedes
from the surface it becomes more compact, the air-
holes being elongated and more compressed. It is not
a little remarkable, that there now exists no trace of
any elevation which could have kept up a lake sufficient to account for such a formation.

On the 8th of May Mr Playfair quitted Rome,
and returned to Florence, following the western road
by Viterbo and Sienna. From Rome to the river
Paglia, a distance of ninety miles, excepting in the
vicinity of the lake of Bolsena, nothing is to be seen
but volcanic tufa, containing lava in large rolled
masses, studded with leucites. Among the beautiful woods which surround that lake, the lava covers

the whole surface, assuming, for a considerable extent, the form of regular basaltic columns in every variety of position, from the horizontal to the vertical : these are very hard, with a conchoidal fracture, and abound in crystals of leucite and pyroxene.   A little beyond the lake at San Lorenzo, although the internal structure remains the same, the columnar form is lost in the appearance of a common stream of lava, but occurs again at Acquapendente, beyond which the volcanic country ceases to be well defined, the tufa being mixed with the marly earth of the hills.   After traversing a plain of this description, the road ascends the high hill of Radicofani, which is crowned by a perpendicular rock of an equivocal character, but more nearly allied to lava than to greenstone.   Some specimens of it are very compact, others are very porous, and all around are scattered scoriæ, resembling those of Vesuvius, mixed with stones of a red colour, full of vesicular cavities, and so light as to float in water.

Here the volcanic country terminates, and is succeeded by a tender marly schistus, interstratified with a feeble sandstone, in thin horizontal beds, presenting a sterile surface, cut into a succession of deep ravines and muddy scars.   Beyond Sienna this gives place to limestone, which, with the exception of a partial formation of travertine, continues to Florence. The travertine extends for four miles along the sides of a valley, at a height of 200 feet above the present

bed of the river, by which it is still deposited, and
then suddenly terminates at a point where the hills,
approaching so closely as almost to shut up the valley, indicate the former existence of a barrier, which
might have kept up a lake sufficient to account for
the formation of this singular substance.

From Florence Mr Playfair proceeded by Lucca,
Pisa, and the Riviera di Levante, to Genoa, which,
after a few days, he left for Turin.   The chief object of curiosity in this route was the marble of Carrara, of which the quarries are situated about three
miles from that city, in a wild desolate valley on the
banks of the Carione.   The strata are highly inclined, and intersected by numerous fissures, which
cut one another at the angle peculiar to the crystallization of the carbonate of lime.   The fine white
saline marble, employed in the arts, lies between
beds of blue limestone, of no value, and that are
removed by gunpowder; but the whole process
is exceedingly slovenly, and no pains have been taken even to make a tolerable road from the quarries
to the sea-shore.

In prosecuting his journey from Turin, by Milan,
to Venice, he traversed in its full extent the great
plain of Lombardy, in which nothing is more remarkable than the enormous quantity of water-worn
gravel, composed of gneiss, granite, porphyry, puddingstone, and limestone.   In the vicinity of the
mountains it forms at least one-third of the soil, di-

minishing in size and in quantity towards the centre
of the level ground.   While it rises occasionally to
a considerable height, as at Peschiera, between the
vallies of the Mincio and the Adige, where it forms a
ridge five hundred feet high, the depth to which it
extends has never been ascertained, the wells being
always filled with water before they reach down to any
rock.   There is, however, a distinct line, at no greater distance from the surface than three feet, above
which it is all red, and below an unmixed grey.

From Montebello he made an excursion to the
hills of the Vicentine territory, which form a low
range extending from Montecchio Maggiore to Cas-
tel Gomberto.   At the first of these places, three
miles from Montebello, rises a hill, one end of which
is limestone, and the other a mass of amygdaloid,
containing calcareous spar, analcime, and calcedony.
In the immediate vicinity is a quarry of a hard compact black basalt, showing a strong tendency to the
columnar form in the interior; and at Brendola, a
mile and a half to the south, is another quarry of a
hard compact greenstone, much more tough than
any ordinary lava.   To the same class belongs the
Monte Berico, which rises from the plain close to
Vicenza, and is composed of limestone, with an
amygdaloid at the summit, while the Monte Tondo,
a smaller elevation attached to its base, consists of
trap tuff, going fast to decay, and including nuclei
of a limestone, which has a smooth conchoidal frac-

ture, and is much more dense and compact than that of the Monte Berico itself. He was prevented by the intense heat (it being now the end of June) from visiting the Euganean mountains, which stand insulated in the level plain ; but in the ample and excellent collection of the Conte del Rio, at Padua, he had an opportunity of seeing their productions, and received from the Count himself the most accurate information respecting the relations of the different rocks to one another. The rock which constitutes the main body of the Euganean hills is strikingly like the whitish lava of Epomeo in Ischia, and many of the specimens contained crystals of feldspath, exactly like those of the lava of the Zolfatara. Some specimens taken from what has the appearance of a current of lava running down the side of one of the hills, are compact and dense in the centre, light, porous, and vesicular at the surface. These hills also abound in pitchstone, beautiful pearlstone, and various approximations to volcanic glass. From all these circumstances, so very different from the hills of the Vicentine, it seems probable that the Euganean mountains are truly volcanic.

On quitting Venice, Mr Playfair entered the Alps by a road which, following the Brenta to its source, in the Valsugana, and then descending the valley of the Pergine, terminates at Trent, in the Tyrol. From Trent to Inspruck the road passes successively through the vallies of the Adige, of the Eisach,

and of the Sihl, by which last it is conducted to the
pass of the Brenner, whence it descends rapidly to
Inspruck.   At a short distance above Trent the valley is suddenly contracted, and reduced, for a distance of twenty miles, to little more than a defile,
which the Adige has cut through an immense wall
of porphyry, which here crosses its course.   This
porphyry, so abundant in the gravel of Lombardy,
and so seldom to be met with in its native place,
has a purple ground, spotted with white feldspath,
and hexagonal crystals of very transparent quartz.

From Inspruck he directed his course through
Bavaria to Lindau, on the lake of Constance, with
the intention of examining the part of Switzerland
which the heavy rains had prevented him from visiting in the preceding summer.   From Lindau he
ascended the valley of the Upper Rhine to Coire,
in the Grisons, and after returning as far as Ma-
lans to cross the river, proceeded by the lakes of
Wallenstatt and Zurich to Lucerne.   From Lucerne
he made an excursion of fourteen days, in which
he traversed the most interesting portion of the
Alps.   After visiting the summit of Rigi, as affording an extensive view of the country he was about
to enter upon, he proceeded by Schwytz, Altorf, and
Andermatt, across the pass of St Gothard to Airolo, in the Val Bedretto, and then crossing the elevated pass of Nufenen, (7336 feet above the level
of the sea,) returned by the valley of Eginen to

Obergesteln, in the Valais; whence, after passing
the Grimsel, he followed the course of the Aar to
Meyringhen, and having visited the vallies of Grin-
delwald and Lauterbrun, returned by Unterseen,
Brientz, and Alpnach, to Lucerne.

While in Switzerland, Mr Playfair made very few
notes, except as references to the works of Saussure
and of Ebel, which he carried as his guides; and
we have, therefore, merely pointed out the line he
followed, as one well adapted for the examination of
that interesting country. Of one object, however,
the slide by which the trees of Pilatus are conveyed
into the lake of Lucerne, he has left a description,
which, although not sufficiently perfect to be published as a separate essay, is inserted in the Appendix,
as too valuable to be lost to the world.

On quitting Lucerne, Mr Playfair again visited
Geneva, where he made a short stay to enjoy the society of his friends, and then proceeded to Lyons on
his way home. We have already observed that his
Notes on Switzerland contain few particular observations; but the following fragment of a paper, drawn
up during his residence at Geneva, will show the views
he had taken of the general structure of the country.

" Switzerland is not more remarkable for its picturesque and sublime scenery, than for the instructive lessons that may be derived from its mountains,
its rivers, and its plains. The mountains are the

highest in Europe, its rivers among the greatest and most rapid, and its plains, if not the most extensive, are certainly those in which the changes they have undergone are most distinctly recorded.

" The mountains may be distinguished into three classes or orders ; the first consisting of those which raise their heads above the circle of perpetual congelation ; the last of the hills or mountains which are clothed to their summits with wood, and the second or intermediate, are those which lie between the other two. The lines, it is evident, which divide these orders, are not arbitrary nor imaginary, but are precisely drawn by the hand of Nature herself. They are lines by which the mountains of every climate, if they reach beyond the circle of perpetual frost, must necessarily be divided into three distinct orders.

" The general arrangement of the rocks, or of the soil, of which the country consists, is very simple. The central chains, those, namely, which bound the Valais on the north and on the south, uniting together in St Gothard, diverge from thence toward the north-east, including the upper part of the Rhine between them, as they had before included the upper part of the Rhone. These chains are in their highest parts of granite or of syenite. If we descend from this great barrier toward the north, we come soon to the calcareous mountains, of which many rise above the circle of perpetual frost. The calcareous mountains comprehend, therefore, many mountains of the first order, and with a few exceptions, all those of the second. At the foot of this calcareous formation, which is primary, or of transition, is stretched out a great mass of puddingstone, over the whole length of the mountainous chain ; sometimes rising into mountains, as in the case of Rigi, and along with this, though exterior and above it, is an equally extensive formation of gres, or sandstone, in horizontal beds, for the most part, and extending north-east and south-west from the Rhine at the Lake of Constance, to the Rhone where it issues from the Lake of Geneva. This sandstone is, in some places, a stone of excellent quality for building, as at St Gall, at Lucerne, at Berne, and at Geneva. When of this quality, it resembles exceedingly the extensive formation that lies at the foot of the Grampians in Scotland, and reaches from the German Ocean nearly to the shores of the Atlantic. Both are characterised by containing little quartz, but much sand, composed of feldspath, mica, and hornblende. It seems as if made up of porphyry, that had been reduced to sand, and it is nearly akin to grauwakke. Exterior to all this, still receding from the mountains, is a quantity of loose gravel, which covers the rock, whatever it be, to a great depth. This gravel is of great superficial extent, and conducts us to the mountains of Jura, on the west and on the north to the Rhine and beyond it. At Jura, a calcareous formation emerges from under the gravel, the sandstone, or the puddingstone, and forms the whole of the great chain to which the name just mentioned is usually applied.

" From the great chain which is properly the Alps, the ground, as far as the Rhine or Jura, though without mountains, rises into hills and ridges covered with wood, and intersected by the great vallies in which the rivers flow, forming altogether a great extent of cultivated land, or of picturesque scenery, such as could hardly be exceeded.

" From the great lakes which are *enclavés*, as it were, in the mountains, descend the great rivers which traverse the plains of Switzerland, and which fall into the Rhine. Of these the most easterly is the Upper Rhine itself ; and next coming westward is the Limmat, which issues from the Lake of Zurich. After this the Reuss, which, descending from St Gothard, and purifying its waters in the Lake of Lucerne, issues from thence a great river, which, as well as the Limmat, joins the Aar before it falls into the Rhine. The Aar comes from the Lake of Thun, and being joined by a large supply from the Lakes of Neuchâtel and Bienne, from the west, as well as the rivers just mentioned from the east, unites its waters to the Rhine itself at Zursach.

" This great system of rivers has every where left traces of the changes it has been the instrument of producing, and indications, which it is impossible to

mistake, of the vast difference between the surface
as it exists at present, and as it has existed in former ages.

" The vast quantity of gravel that occupies nearly
all the low country of Switzerland, has already been
remarked. The ground is never opened in any
place, nor an abrupt face found on the banks of a
river, where it does not appear that the whole consists of gravel and sand. The gravel consists of the
stones which belong to the high Alps, granite, gneiss,
mica slate, hornblende schistus, petrosilex, jade,
hard limestone, &c. The figure of the stones is
worked remarkably true, and the polish in general
is fine. The consequence of the rivers running
through such materials, has been the formation of
terraces on a vast scale, and in great numbers. In
many places, a succession of such terraces, as far as
five, may be counted ; as far as two and three are
very common. The height of one above another is
from twenty to thirty, and even forty feet. At the
Rhine they are very conspicuous ; one, of which I
measured the perpendicular height, was 122 feet
above the present surface of the river. When it is
considered that three or even four of such terraces
can often be counted on the banks of this great
river, it may fairly be stated, that the evidence of
the Rhine having flowed at the height of 360 feet
above the present level, is very conclusive."******

Even more remarkable than the above are the

terraces of gravel upon which the road through the
Tyrol is conducted from Sterzing to Inspruck, at a
height of 500 feet above the present bed of the
river.

After examining the rocks of Fort St Jean, and of
Pierre Encyze, where the granite veins are distinctly seen issuing from the central mass, and penetrating the strata of mica slate which rest upon its sides,
he proceeded from Lyons to Clermont in Auvergne;
a route which led him across a granite country forming a succession of hills, nowhere rising to any
considerable height, and rounded at their summits.
This granite is of a very perishable nature, and
yields readily to the weather, except where traversed
by dikes of greenstone, which occur frequently, and
are often of a great size. One of them, cut across
by the road near Fenouilh, is forty feet wide, and
firmly united at the edges with the granite, which
is considerably indurated at the point of contact.
On approaching Clermont, which is situated in the
valley of the Allier, the granite is concealed by the
alluvial earth, but the greenstone is still visible in
large masses, and, at intervals, a limestone rock remarkable for exudations of pitch, containing loose
crystals of calcedony.

The hill nearest to Clermont is Graveneyre, of
which the sides are covered with scoriæ and cinders,
having the same black colour, vitreous fracture, and
scorious surface with those of the Atrio del Cavallo;

and the summit is a flat surface, from which rises, on one side, a red face of rock resembling very exactly the remains of the old craters of Vesuvius. Two streams of lava can be distinctly traced, one terminating at the bottom of the hill towards Clermont, and the other running down to the valley of Royat, which it follows for the remainder of its course. The surface of both is scorious, while, in the valley just mentioned, the stream is seen equally scorious above and below, and resting upon alluvial earth. There can be no doubt, then, that this is a true lava : and Graveneyre must be an extinguished volcano of a very recent date ; for every thing indicates that the valley of Royat, when the lava first entered it, was nearly the same as it is at the present day. The insulated conical hills, known by the generic name of Puy, rise from a plain or table land of granite, tolerably level, but elevated more than 400 feet above the flat country in which Clermont is situated. Of these the Puy de Pariou affords the most perfect specimen. It is a cone covered with fine turf, both on the ascent and within the crater, which is a mile in circumference, and very deep, sloping downwards at an angle of 30°. From the lower part of the cone issues a considerable current of lava, still rugged and black, and the plain is covered with scoriæ and volcanic cinders, which are seen to the depth of twenty feet in the cuts made by the winter rains. Near the extremity of this current of lava, but totally unconnected with it, is a range of basaltic columns, to which it is not easy to assign a place.   They are very regular, have four and five sides from six to nine inches wide, and are in close contact with each other.   The internal structure is very singular; a number of blue spherulae of the size of a pea, are firmly united by a brown cement, forming a stone more compact than any lava to be seen in this country, and yet less so than trap. There is no appearance of scoriousness in any part of the mass, and they rest immediately upon the granite already mentioned as the basis of the whole plain.

In order to obtain a more general knowledge of this curious range of hills, Mr Playfair rode down the valley of the Allier, at the foot of the granite plain, to Volvic, whence he followed a current of lava to its source, ascended the Puy Chopine, and returned to Clermont by the Puy de Pariou, already described.   Between Clermont and Volvic several streams of lava descend into the plain, but none to be compared, as to extent, with that which bears the name of Volvic.   It is three miles in length, a mile in breadth, and the surface is every where scorious and rugged in the extreme.   The rock itself, as seen in the extensive quarries, is of a greyish blue colour, porous, splintery in the fracture, and contains only a few crystals of pyroxene.   It is raised by wedges, and so easily worked by the chisel, as to be used in

the vicinity for building, and sent to Paris, where it
is bored into pipes for the conveyance of water.   It
is only towards the centre that it is sufficiently compact to be applied to these purposes, the remainder,
both above and below, being vesicular and scorious.
This stream can be traced to its source in the side
of the Puy de la Nugère, within a short distance of
which rises the Puy Chopine, the most interesting,
as well as the most remarkable of the whole range.
The western side is laid open nearly from top to
bottom by a large scar, which shows the rock to be
a lava closely resembling the greyish white lava of
the Zolfatara, and, like it, containing crystals of
feldspath.   The northern edge of this scar, covered
with grass and bushes, affords an easy, though steep
ascent to the summit, which forms an irregular crater of small depth.   The eastern side is laid bare
like the western, but the rock is the small grained
granite of the plain.   In one place only, nearly half
way down, the junction is visible, and here the lava
and the granite were found firmly united in the same
specimen.

Farther observation showed only a repetition of
the same phenomena; conical hills, formed of scoriæ, with a crater in the centre, and with streams of
lava issuing from their sides, mingled with other
hills formed of the white lava, containing feldspath,
and united with the unaltered granite.   Sarcouy
alone presents the rock altered in structure, raised

up like a bell, but nowhere burst open, and free
from scoriæ and lava. It seems not a little extraordinary, that any one who had examined this singular chain of hills, should ever have doubted their
volcanic origin.

From Clermont Mr Playfair proceeded to Paris,
where he remained for a month, and then returned
home to resume his academical studies, after an absence of seventeen months, and a journey of 4000
miles.

In the ensuing summer, after witnessing the commencement of an Observatory for the Astronomical
Institution, of which he was president, he retired to
Burntisland, to complete the materials of his Dissertation on Mathematical and Physical Science. At
this time, also, was drawn up the Memoir on Na-
val Tactics, published, after his death, in the Transactions of the Royal Society of Edinburgh, which is,
however, only a fragment of a life of Mr Clerk of
Eldin, to whom the world is indebted for the discovery of that valuable system.

It was his intention to publish, in the form of detached papers, an account of the more remarkable
objects which he had met with in his journey, and, after completing the Dissertation, to commence the long
projected edition of the Illustrations of the Huttonian
Theory of the Earth ; but a severe attack of a disease in the bladder, which had long given him occasional uneasiness, interrupted both his lectures and

his studies during a great part of the winter. He, however, regained health and strength sufficient to resume and finish the course of lectures ; but in June the disease recurred with increased violence, and, after an illness of a month, terminated his existence on the 19th of July 1819. Although suffering very severe pain, with very short intervals of rest, he employed himself, until a few days before his death, in dictating corrections on the proof sheets of the Dissertation ; and, even after his bodily strength was exhausted, he retained his intellectual faculties unimpaired to the very last.

IT has struck many people, we believe, as very extraordinary, that so eminent a person as Mr Playfair should have been allowed to sink into his grave in the midst of us, without calling forth almost so much as an attempt to commemorate his merit, even in a common newspaper ; and that the death of a man so celebrated and so beloved, and, at the same time, so closely connected with many who could well appreciate and suitably describe his excellences, should be left to the brief and ordinary notice of the daily obituary. No event of the kind certainly ever excited more general sympathy ; and no individual, we are persuaded, will be longer or more affection-

4

ately remembered by all the classes of his fellow-citizens : and yet it is to these very circumstances that we must look for an explanation of the apparent neglect by which his memory has been followed. His humbler admirers have been deterred from expressing their sentiments by a natural feeling of unwillingness to encroach on the privilege of those whom a nearer approach to his person and talents rendered more worthy to speak of them,—while the learned and eloquent among his friends have trusted to each other for the performance of a task which they could not but feel to be painful in itself, and not a little difficult to perform as it ought to be ; or perhaps have reserved for some more solemn occasion that tribute for which the public impatience is already at its height.

We beg leave to assure our readers that it is merely from anxiety to do *something* to gratify this natural impatience that we presume to enter at all upon a subject to which we are perfectly aware that we are incapable of doing justice : For of Mr Playfair's scientific attainments,—of his proficiency in those studies to which he was peculiarly devoted, we are but slenderly qualified to judge : But, we believe we hazard nothing in saying that he was one of the most learned mathematicians of his age, and among the first, if not the very first, who introduced the beautiful discoveries of the later continental geometers to the knowledge of his countrymen, and gave

their just value and true place, in the scheme of
European knowledge, to those important improvements by which the whole aspect of the abstract
sciences has been renovated since the days of our illustrious Newton.   If he did not signalize himself
by any brilliant or original invention, he must, at
least, be allowed to have been a most generous and
intelligent judge of the achievements of others, as
well as the most eloquent expounder of that great
and magnificent system of knowledge which has been
gradually evolved by the successive labours of so
many gifted individuals.   He possessed, indeed, in
the highest degree, all the characteristics both of a
fine and a powerful understanding,—at once penetrating and vigilant,—but more distinguished, perhaps, for the caution and sureness of its march, than
for the brilliancy or rapidity of its movements,—and
guided and adorned through all its progress by the
most genuine enthusiasm for all that is grand, and
the justest taste for all that is beautiful in the Truth
or the Intellectual Energy with which he was habi-
tually conversant.

To what account these rare qualities might have
been turned, and what more brilliant or lasting
fruits they might have produced, if his whole life
had been dedicated to the solitary cultivation of
science, it is not for us to conjecture ; but it cannot
be doubted that they added incalculably to his eminence and utility as a teacher ; both by enabling

him to direct his pupils to the most simple and luminous methods of inquiry, and to imbue their
minds, from the very commencement of the study,
with that fine relish for the truths it disclosed, and
that high sense of the majesty with which they were
invested, that predominated in his own bosom.
While he left nothing unexplained or unreduced to
its proper place in the system, he took care that
they should never be perplexed by petty difficulties,
or bewildered in useless details, and formed them
betimes to that clear, masculine, and direct method
of investigation, by which, with the least labour, the
greatest advances might be accomplished.

Mr Playfair, however, was not merely a teacher;
and has fortunately left behind him a variety of
works, from which other generations may be enabled to judge of some of those qualifications which
so powerfully recommended and endeared him to
his contemporaries.   It is, perhaps, to be regretted
that so much of his time, and so large a proportion
of his publications, should have been devoted to the
subjects of the Indian Astronomy, and the Huttonian Theory of the Earth.   For though nothing can
be more beautiful or instructive than his speculations on those curious topics, it cannot be dissem-
bled that their results are less conclusive and satisfactory than might have been desired ; and that his
doctrines, from the very nature of the subjects, are
more questionable than we believe they could pos-

sibly have been on any other topic in the whole circle of the sciences.   To the first, indeed, he came
under the great disadvantage of being unacquainted
with the Eastern tongues, and without the means of
judging of the authenticity of the documents which
he was obliged to assume as the elements of his reasonings; * and as to the other, though he ended,
we believe, with being a very able and skilful mineralogist, we think it is now generally admitted that
that science does not yet afford sufficient materials
for any positive conclusion; and that all attempts
to establish a Theory of the Earth must, for many
years to come, be regarded as premature.   Though
it is impossible, therefore, to think too highly of the
ingenuity, the vigour, and the eloquence of those
publications, we are of opinion that a juster estimate
of Mr Playfair's talent, and a truer picture of his
genius and understanding, is to be found in his
other writings;—in the papers, both biographical
and scientific, with which he has enriched the Transactions of our Royal Society;—his account of Laplace, and other articles which he is understood to

---

* The authenticity of the Indian tables is inferred, not so
much from the history attached to them, as from the accuracy
with which they describe the celestial phenomena of the period to which they refer.   No one but an astronomer acquainted with the latest refinements of European science,
could have produced such a work by calculating from the present state of the heavens.   *Vide* Vol. III. p. 112, *et seq.*—Ed.

have contributed to the Edinburgh Review,—the Outlines of his Lectures on Natural Philosophy,— and, above all, his Introductory Discourse to the Supplement to the Encyclopædia Britannica, with the final correction of which he was occupied up to the last moments that the progress of his disease allowed him to dedicate to any intellectual exertion.

With reference to these works, we do not think we are influenced by any national, or other partiality, when we say that he was certainly one of the best writers of his age ; and even that we do not now recollect any one of his contemporaries who was so great a master of composition.   There is a certain mellowness and richness about his style, which adorns, without disguising the weight and nervousness, which is its other great characteristic,—a sedate gracefulness and manly simplicity in the more level passages,—and a mild majesty and considerate enthusiasm where he rises above them, of which we scarcely know where to find any other example. There is great equability, too, and sustained force in every part of his writings.   He never exhausts himself in flashes and epigrams, nor languishes into tameness or insipidity ; at first sight you would say that plainness and good sense were the predominating qualities ; but by and bye, this simplicity is enriched with the delicate and vivid colours of a fine imagination,—the free and forcible touches of a most powerful intellect,—and the lights and shades of an unerring and harmonizing taste. In comparing it with the styles of his most celebrated contemporaries, we would say that it was more purely and peculiarly a *written* style,—and, therefore, rejected those ornaments that more properly belong to oratory. It had no impetuosity, hurry, or vehemence,—no bursts or sudden turns or abruptions, like that of Burke ; and though eminently smooth and melodious, it was not modulated to an uniform system of solemn declamation like that of Johnson, nor spread out in the richer and more voluminous elocution of Stewart ; nor still less broken into that patch-work of scholastic pedantry and conversational smartness which has found its admirers in Gibbon. It is a style, in short, of great freedom, force, and beauty ; but the deliberate style of a man of thought and of learning, and neither that of a wit throwing out his extempores with an affectation of careless grace,—nor of a rhetorician thinking more of his manner than his matter, and determined to be admired for his expression, whatever may be the fate of his sentiments.

His habits of composition, as we have understood, were not perhaps exactly what might have been expected from their results. He wrote rather slowly, —and his first sketches were often very slight and imperfect,—like the rude chalking for a masterly picture. His chief effort and greatest pleasure was in their revisal and correction ; and there were no limits to the improvement which resulted from this

application.   It was not the style merely, or indeed
chiefly, that gained by it : The whole reasoning,
and sentiment, and illustration, was enlarged and
new modelled in the course of it, and a naked outline became gradually informed with life, colour,
and expression.   It was not at all like the common finishing and polishing to which careful authors generally subject the first draughts of their
compositions,—nor even like the fastidious and ten-
tative alterations with which some more anxious writers assay their choicer passages.   It was, in fact,
the great filling in of the picture,—the working up
of the figured *weft*, on the naked and meagre *woof*
that had been stretched to receive it; and the singular thing in his case was, not only that he left
this most material part of his work to be performed
after the whole outline had been finished, but that
he could proceed with it to an indefinite extent, and
enrich and improve as long as he thought fit, without any risk either of destroying the proportions of
that outline, or injuring the harmony and unity of
the design.   He was perfectly aware, too, of the
possession of this extraordinary power, and it was
partly, we presume, in consequence of it that he
was not only at all times ready to go on with any
work in which he was engaged, without waiting for
favourable moments or hours of greater alacrity, but
that he never felt any of those doubts and misgiv-
ings as to his being able to get creditably through

with his undertaking, to which we believe most authors are occasionally: liable. As he never wrote upon any subject of which he was not perfectly master, he was secure against all blunders in the substance of what he had to say ; and felt quite assured, that if he was only allowed time enough, he should finally come to say it in the very best way of which he was capable. He had no anxiety, therefore, either in undertaking or proceeding with his tasks ; and intermitted and resumed them at his convenience, with the comfortable certainty, that all the time he bestowed on them was turned to good account, and that what was left imperfect at one sitting might be finished with equal ease and advantage at another. Being thus perfectly sure both of his end and his means, he experienced, in the course of his compositions, none of that little fever of the spirits with which that operation is so apt to be accompanied. He had no capricious visitings of fancy which it was necessary to fix on the spot or to lose for ever,—no casual inspirations to invoke and to wait for,—no transitory and evanescent lights to catch before they faded. All that was in his mind was subject to his control, and amenable to his call, though it might not obey at the moment ; and while his taste was so sure, that he was in no danger of over-working any thing that he had designed, all his thoughts and sentiments had that unity and congruity, that they fell almost spontaneously into

harmony and order; and the last added, incorporated, and assimilated with the first, as if they had
sprung simultaneously from the same happy con-
ception.

But we need dwell no longer on qualities that
may be gathered hereafter from the works he has
left behind him. They who lived with him mourn
the most for those which will be traced in no such
memorial; and prize far above those talents which
gained him his high name in philosophy, that personal character which endeared him to his friends,
and shed a grace and a dignity over all the society in
which he moved. The same admirable taste which
is conspicuous in his writings, or rather the higher
principles from which that taste was but an emana-
tion, spread a similar charm over his whole life and
conversation; and gave to the most learned philosopher of his day the manners and deportment of the
most perfect gentleman. Nor was this in him the
result merely of good sense and good temper, assisted by an early familiarity with good company, and a
consequent knowledge of his own place and that of
all around him. His good breeding was of a higher
descent; and his powers of pleasing rested on something better than mere companionable qualities.
With the greatest kindness and generosity of nature, he united the most manly firmness, and the
highest principles of honour,—and the most cheerful and social dispositions, with the gentlest and

steadiest affections. Towards women he had always
the most chivalrous feelings of regard and attention,
and was, beyond almost all men, acceptable and
agreeable in their society,—though without the least
levity or pretension unbecoming his age or condition : And such, indeed, was the fascination of the
perfect simplicity and mildness of his manners, that
the same tone and deportment seemed equally appropriate in all societies, and enabled him to delight
the young and the gay with the same sort of conversation which instructed the learned and the grave.
There never, indeed, was a man of learning and talent who appeared in society so perfectly free from
all sorts of pretension or notion of his own importance, or so little solicitous to distinguish himself,
or so sincerely willing to give place to every one
else. Even upon subjects which he had thorough-
ly studied, he was never in the least impatient to
speak, and spoke at all times without any tone of
authority ; while, so far from wishing to set off what
he had to say by any brilliancy or emphasis of expression, it seemed generally as if he had studied to
disguise the weight and originality of his thoughts
under the plainest form of speech and the most quiet
and indifferent manner : so that the profoundest remarks and subtlest observations were often dropped,
not only without any solicitude that their value
should be observed, but without any apparent con-
sciousness that they possessed any. Though the

most social of human beings, and the most disposed
to encourage and sympathize with the gaiety and
joviality of others, his own spirits were in general
rather cheerful than gay, or at least never rose to
any turbulence or tumult of merriment; and while
he would listen with the kindest indulgence to the
more extravagant sallies of his younger friends, and
prompt them by the heartiest approbation, his own
satisfaction might generally be traced in a slow and
temperate smile, gradually mantling over his bene-
volent and intelligent features, and lighting up the
countenance of the Sage with the expression of
the mildest and most genuine philanthropy.   It
was wonderful, indeed, considering the measure
of his own intellect, and the rigid and undeviat-
ing propriety of his own conduct, how tolerant
he was of the defects and errors of other men.
He was too indulgent, in truth, and favourable
to his friends;—and made a kind and liberal allowance for the faults of all mankind—,except
only faults of baseness or of cruelty,—against which
he never failed to manifest the most open scorn
and detestation.   Independent, in short, of his high
attainments, Mr Playfair was one of the most
amiable and estimable of men,—delightful in his
manners,—inflexible in his principles, and generous
in his affections, he had all that could charm in society or attach in private; and while his friends en-

joyed the free and unstudied conversation of an easy and intelligent associate, they had at all times the proud and inward assurance that he was a being upon whose perfect honour and generosity they might rely with the most implicit confidence, in life and in death,—and of whom it was equally impossible, that, under any circumstances, he should ever perform a mean, a selfish, or a *questionable* action, as that his body should cease to gravitate or his soul to live !

If we do not greatly deceive ourselves, there is nothing here of exaggeration or partial feeling,—and nothing with which an indifferent and honest chronicler would not concur. Nor is it altogether idle to have dwelt so long on the personal character of this distinguished individual : For we are ourselves persuaded, that this personal character has almost done as much for the cause of science and philosophy among us as the great talents and attainments with which it was combined,—and has contributed in a very eminent degree to give to the better society of this our city that tone of intelligence and liberality by which it is so honourably distinguished. It is not a little advantageous to philosophy that it is in fashion,—and it is still more advantageous, perhaps, to the society which is led to confer on it this apparently trivial distinction. It is a great thing for the country at large,—for its happiness, its prosperity, and its renown,—that the upper and influen-

cing part of its population should be made familiar,
even in its untasked and social hours, with sound and
liberal information, and be taught to know and respect those who have distinguished themselves for
great intellectual attainments.   Nor is it, after all, a
slight or despicable reward for a man of genius to be
received with honour in the highest and most ele-
gant society around him, and to receive in his living
person that homage and applause which is too often
reserved for his memory.   Now, those desirable ends
can never be effectually accomplished, unless the
manners of our leading philosophers are agreeable,
and their personal habits and dispositions engaging
and amiable.   From the time of Hume and Robertson, we have been fortunate in Edinburgh in possessing a succession of distinguished men, who have
kept up this salutary connection between the learned
and the fashionable world ; but there never, perhaps,
was any one who contributed so powerfully to confirm and extend it, and that in times when it was
peculiarly difficult, as the lamented individual of
whom we are now speaking ; and they who have had
the most opportunity to observe how superior the society of Edinburgh is to that of most other places of
the same size, and how much of that superiority is
owing to the cordial combination of the two aristo-
cracies, of rank and of letters,—of both of which it
happens to be the chief provincial seat,—will be best

able to judge of the importance of the service he has thus rendered to its inhabitants, and through them, and by their example, to all the rest of the country.

APPENDIX.

No. I.

JOURNAL, &c.

" HAVING obtained leave of absence for some months in the beginning of the year 1782, I determined to visit the metropolis, that I might have an opportunity of seeing what is there most worthy of observation, and of conversing with those men whose names are known in the republic of letters. This last indeed was my principal object, and I accordingly put down those passages in conversation, and those circumstances in the characters of the men I saw, that seemed to me most worthy of being remembered. These I have now brought together and connected in the following pages.

" My first care on my arrival was to wait on Dr Maskelyne, the Astronomer Royal, with whom I had become acquainted some years before while he was engaged in his experiments on Schehallien in Perthshire, which have since acquired him so much reputation. I met with a very cordial welcome from him, and found that an acquaintance contract-

ed among wilds and mountains is much more likely
to be durable than one made up in the bustle of a
great city ; nor would I by living in London for
many years have become so well acquainted with
this astronomer as I did by partaking of his hard-
ships and labours on Schehallien for a few days.
Dr Maskelyne is of a middle age, and was preferred
to the honourable station he now fills from his merits only.   He is an excellent observer, and a good
mathematician.   He is much attached to the study
of geometry, and I am not sure that he is very deeply versed in the late discoveries of the foreign alge-
braists.   Indeed, this seems to be somewhat the
case with all the English mathematicians ; they despise their brethren on the Continent, and think that
every thing great in science must be for ever confined to the country that produced Sir Isaac Newton.
Dr Maskelyne, however, is more than almost any of
them superior to this prejudice. He is slow in appre-
hending new truths, but his mind takes a very firm
hold of them at last. He has been of great service to
the art of navigation, by facilitating the method of
discovering the longitude at sea by observations of
the distance of the moon from a fixed star.   But
for the ingenuity and industry of Dr Maskelyne
that method would have remained impracticable to
all but astronomers. Dr Maskelyne has been ac-
cused of sometimes detracting from the discoveries
of others when they interfered with his own ; I must

l

say, however, that I never could observe any thing of this kind, though I saw him placed in one of those critical situations where envy and jealousy, had they lurked anywhere within him, could scarcely have failed to make their appearance.

" At Dr Maskelyne's, soon after my arrival, I was introduced to Dr Horsley. That gentleman, from his papers in the Philosophical Transactions, and his Commentary on the Principia, is considered as being at the head of the English mathematicians. He is a man of abilities ; and his conversation, when the first stiffness is worn off, becomes very pleasant. Our conversation turned first on Lord Monboddo, who is a great friend of Horsley. He expressed great respect for Lord M. for his learning and his acuteness, and (what was more surprising) for the soundness of his judgment. He talked very seriously of the notion of mind being united to all the parts of matter, and being the cause of motion. So far as I could gather, Dr Horsley supposes that every atom of matter has a soul, which is the cause of its motion, its gravitation, &c. What has made him adopt this strange unphilosophical notion I cannot tell, unless it be the fear that his study of natural philosophy should make him suspected of atheism, or at least of materialism. For it is certain that there is at present a prejudice among the English clergy that natural philosophy has a tendency to make men atheists or materialists. This ab-

surd prejudice was first introduced, I think, by
that illiberal though learned prelate, Dr Warbur-
ton.

" This was the first time that I had seen the Observatory of Greenwich, and I entered with profound
reverence into that temple of science, where Flam-
stead, and Halley, and Bradley, devoted their days
and their nights to the contemplation of the heavens.
The shades of these ancient sages seemed still to
hover round their former mansion, inspiring their
worthy successor with the love of wisdom, and
pointing out the road to immortality.

" Though the climate of Greenwich be not very
favourable to observation, yet, such has been the industry of the astronomers belonging to that observatory, that more good observations have been made
there than in any other part of the world.   So
much do the moral causes sometimes control the
physical.   The place of Astronomer Royal has a salary of L. 200 a-year.   Queen Caroline offered to
Dr Halley, who was then Astronomer Royal, an
augmentation of his salary; but that philosopher,
with the disinterestedness of a true lover of science,
declined accepting it, because, he said, while the salary was small, the place would never be an object
to any but an astronomer; should it become more
considerable, it would be sought after for the sake
of emolument, and might be given away from poli-
tical intrigue.   He, therefore, requested of her Ma-

jesty to mark her zeal for science rather by improving the instruments of the observatory, than by augmenting the salary of the astronomer.

" My next care was to visit the British Museum, and to deliver to Dr Solander a letter of introduction which I had brought with me from Dr Robertson. Of the immense collection of natural curiosities, and of historical monuments contained in the museum, it is impossible to speak; a stranger regrets that he has not time to derive any advantage from them, surrounded, as he probably is, with a crowd of ignorant people, and hurried through by guides impatient of the torture which they continually suffer from the impertinence of their guests.

" The good humour of Dr Solander is alone proof against all these assaults of impertinence and folly, and he has never been known to utter an impatient expression, for all the penance that the frivolity of the gay, or the stupidity of the dull, could inflict. He is, indeed, a very pleasant man, has lived much in the world, both of literature and of fashion, and has conversed much both with the polite and the savage. There can be no doubt of his skill as a natural historian, yet I very much doubt if, in the branch of mineralogy, he be very profound. This I say from his recommending to me Linnæus's History of Fossils as the best rudiments of mineralogy. Now, it is certain that that book contains nothing but names and external characters,

and that Linnæus himself was not sufficiently a chemist to understand the theory of the fossil kingdom.
The same, perhaps, is the case with Dr Solander.
But one thing for which I admire him is, that he
takes an interest in all the sciences, and is not of the
number of those naturalists who, while they count the
scales of a salmon, or inspect the wing of a butter-
fly, despise the labours of the moralist or the astronomer.

" I was carried by Dr Solander to dine with the
club of the Royal Society at the Crown and An-
chor.   Though I met here with many people whom
I wished much to see, yet I could not help remarking, that there was little pains taken to make the
company very agreeable to a stranger ; and I had
occasion to pity two or three foreigners that I saw
there, who, as well as myself, had sometimes less attention paid to them than their situation required.
However, this club improved much on better acquaintance, and during my stay in London I fre-
quented it very much.   Here, for the first time, I
found some advantage from having written, two
years before this, a paper in the Philosophical Transactions.   I was considered, at least, as a man of
some industry, and perhaps the title of a Dissertation on Impossible Quantities, conveyed to many
people there an idea of depth much beyond the
reality.

" Here I found Mr Smeaton and Mr Aubert,

the latter a very polite man, and a great consolation to a stranger, amid the inattention of the English philosophers. He is of a French family, a great lover of astronomy, and possessed of the best set of astronomical instruments that belongs, perhaps, to any private man.

" Mr Smeaton is a man of excellent understanding, improved more by very extensive experience and observation, than by learning or education. Some mechanical notions concerning force were the occasion of bringing me at this time very well acquainted with him. He was preparing a paper for the Royal Society, in which he proved that there was, in the collision of bodies, a loss of what he calls mechanical power. From the imperfect manner in which Mr Smeaton explains himself on this subject, wherever he has had occasion to treat it, it has been often confounded with the quantity of motion of the Newtonian philosophers, whereas it is in reality the *vis viva* of the foreign mathematicians. He had put his paper into the hands of Mr Cavendish, and he, understanding the thing in the sense I have mentioned, objected to Mr Smeaton's notion, as inconsistent with some of the laws of motion that are the most perfectly established. On first reading Mr Smeaton's paper, I was exactly of Mr Cavendish's opinion ; but in conversing with Mr Smeaton, through the embarrassment of his language, which is very great, I at last got sight of his true

meaning, and made him very happy by assuring
him, that what he said was no way inconsistent with
the Newtonian doctrine of motion. Truth, however, made it necessary for me to tell him, that when
what he said was properly understood, it appeared
to me to be very true, but to be by no means new,
having been often taken notice of by the foreigners
in treating of hard and soft bodies. This I told him
with all the softness I could, imagining that he
might be hurt at it, and knowing well, that to tell
an author his discoveries are not new, is the next
thing to telling him that they are not true. But
Mr Smeaton, luckily for me, did not feel in that
manner; he said, that if it was known, it was so
only to a few of the mathematicians, that by engineers, for whom he chiefly wrote, it was not sufficiently understood, and that his experiment, if not
his conclusion, was a new one.

" Mr Smeaton has much the appearance of an
honest and worthy man; his manners not much polished, but his conversation most instructive in every
thing that relates to mechanics, or the business of
an engineer. He was bred a mathematical instrument maker, and it is to be regretted that an education probably not liberal has deprived him of the
power of becoming deeply versed in mathematics,
and the sublimer parts of natural philosophy.

" Mr Cavendish is a member also of this meeting. He is of an awkward appearance, and has

certainly not much the look of a man of rank. He speaks likewise with great difficulty and hesitation, and very seldom. But the gleams of genius break often through this unpromising exterior. He never speaks at all but that it is exceedingly to the purpose, and either brings some excellent information, or draws some important conclusion. His knowledge is very extensive and very accurate; most of the members of the Royal Society seem to look up to him as to one possessed of talents confessedly superior; and, indeed, they have reason to do so, for Mr Cavendish, so far as I could see, is the only one among them who joins together the knowledge of mathematics, chemistry, and experimental philosophy.

" Chemistry is the *rage* in London at present. I was introduced by Mr B. Vaughan (with whom I became acquainted in Edinburgh while he studied at the university there) to a chemical society, which meets once a fortnight at the Chapter Coffee-house. Here I met Mr Whithurst, a venerable old man, author of an Inquiry into the Formation of the Earth, Dr Keir, Dr Craufurd, and several others. The conversation was purely chemical, and turned on Bergmann's experiments on iron. An anecdote of some Indians was told, that struck me very much, as holding up but too exact a picture of many of our theories and reasonings from analogy. Some American savages having experienced the effects of gunpowder, and having also accidentally become masters of a small quantity of it, set themselves to examine it, with a design of finding out what was its nature, and how it was to be procured. The oldest and wisest of the tribe, after considering it attentively, pronounced it to be a seed. A piece of ground was accordingly prepared for it, and it was sown in the fullest confidence that a great crop of it was to be produced.

" We smile at the mistake of these Indians, and we do not consider, that, for the extent of their experience, they reasoned well, and drew as logical a conclusion as many of the philosophers in Europe. Whenever we reason only from analogy and resemblance, and whenever we attempt to measure the nature of things by our conceptions, we are precisely in the situation of these poor Americans.

" Mr Vaughan and his father are both of them dissenters, and at their house I often found all the chief men of that interest assembled; Dr Price, Priestley, Kippis, Tours, and a number of others. To be a Scotsman was far, I soon found, from being any recommendation to these gentlemen, and they seemed to look on the members of every established church with contempt or abhorrence. The manners of Dr Price were the softest by far of any among them, and I found myself easiest in his company. He is certainly a good mathematician, but politics absorb at present all his thoughts.

Dr Priestley has made so great a figure in the world, that my anxiety to see him was very great. But his conversation has nothing in it very remarkable. When politics are the subject of discourse, he has the same violence with his brethren, and savours not much either of soundness of head or extent of information. On the subjects of chemistry, and the doctrine of fixed air, he talked, indeed, with a great deal of acuteness, and like a man that had been long conversant with experimental philosophy. He was at this time particularly engaged in some experiments, to prove that inflammable air is the same thing with phlogiston. He had revived the calces of several metals, by shutting them up in a receiver with inflammable air, and making the focus of a burning glass to fall on them, they were revived and converted into their respective metals, in the same way as if they had had charcoal added to them when they were exposed to the heat.

This experiment Dr Priestley has since published, but by the best chemists, it is considered as insufficient to prove the point in question. Dr Priestley is very sanguine in the forming of theories, which he does very often, without sufficient data, a fault that is perhaps compensated by the facility with which he afterwards abandons them. On the whole, from Dr Priestley's conversation, and from his writings, one is not much disposed to consider him as a .

person of first rate abilities. The activity, rather than the force of his genius, is the object of admiration. He is indefatigable in making experiments, and he compensates by the number of them, for the unskilfulness with which they are often contrived, and the hastiness with which conclusions are drawn from them. Though little skilled in mathematics, he has written on optics with tolerable success, and though but moderately versed in chemistry, he has done very considerable service to that science.

If we view him as a critic, a metaphysician, and a divine, we must confine ourselves to more scanty praise. In his controversy with Dr Reid, though he has said many things that are true, he has shown himself wholly incapable of understanding the principal point in debate ; and when he has affirmed that the vague and unsatisfactory speculations of Hartley have thrown as much light on the nature of man, as the reasonings of Sir Isaac Newton did on the nature of body, he can scarcely be allowed to understand in what true philosophy consists. As to his theology, it is enough to say that he denies the immateriality of the soul, though he contends for its immortality, and ranges himself on the side of Christianity. These inconsistencies and absurdities will perhaps deprive him of the name of a philosopher, but he will still merit the name of an useful and diligent experimenter."

No. II.

ACCOUNT

OF

THE SLIDE OF ALPNACH.

———

" On the south side of Pilatus, a considerable mountain near Lucerne, are great forests of spruce fir, consisting of the finest timber, but in a situation which the height, the steepness, and the ruggedness of the ground, seemed to render inaccessible. They had rarely been visited but by the chamois hunters, and it was from them, indeed, that the first information concerning the size of the trees and the extent of the forest appears to have been received. These woods are in the canton of Unterwalden, one of those in which the ancient spirit of the Swiss republics is the best preserved ; where the manners are extremely simple, the occupations of the people mostly those of agriculture, where there are no manufactures, little accumulation of capital, and no commercial enterprise. In the possession of such masters, the lofty firs of Pilatus were likely to remain long the ornaments of their native mountain.

" A few years ago, however, Mr Rupp, a native of
Wirtemberg, and a skilful engineer, in which profession he had been educated, indignant at the poli-
tical changes effected in his own country, was induced to take refuge among a free people, and came
to settle in the canton of Schwytz, on the opposite
side of the lake of Lucerne. The accounts which
he heard there of the forest just mentioned determined him to visit it, and he was so much struck
by its appearance, that, long and rugged as the descent was, he conceived the bold project of bringing
down the trees by no other force than their own
weight into the lake of Lucerne, from which the
conveyance to the German Ocean was easy and ex-
peditious. A more accurate survey of the ground
convinced him of the practicability of the project.

" He had by this time resided long enough in
Switzerland to have both his talents and integrity in
such estimation, that he was able to prevail on a
number of the proprietors to form a company, with
a joint stock, to be laid out in the purchase of the
forest, and in the construction of the road along
which it was intended that the trees should slide
down into the lake of Lucerne, an arm or gulf of
which fortunately approaches quite near to the bottom of the mountain. The sum required for this
purpose was very considerable for that country,
amounting to nine or ten thousand pounds; three
thousand to be laid out on the purchase of the fo-

rest, from the community of Alpnach, the proprietors of it, and the rest being necessary for the construction of the singular railway by which the trees
were to be brought down. In a country where
there is little enterprise, few capitalists, and where
he was himself a stranger, this was not the least difficult part of Mr Rupp's undertaking.

"" The distance which the trees had to be conveyed is about three of the leagues of that country, or,
more exactly, 46,000 feet. The medium height of
the forest is about 2500 feet; (which measure I
took from General Pfyffer's model of the Alps, and
not from any actual measurement of my own.) The
horizontal distance just mentioned, when reduced to
English measure, making allowance for the Swiss
foot, is 44,252 feet, eight English miles and about
three furlongs. The declivity is therefore one foot
in 17.68; the medium angle of elevation 3° 14' 20".

"" This declivity, though so moderate, on the
whole, is, in many places, very rapid; at the beginning the inclination is about one-fourth of a right
angle, or about 22° 30'; in many places it is 20°,
but nowhere greater than the angle first mention-
ed, 22° 30'. The inclination continues of this quantity for about 500 feet, after which the way is less
steep, and often considerably circuitous, according
to the directions which the ruggedness of the ground
forces it to take.

"" Along this line the trees descend, in a sort of

trough, built in a cradle form, and extending from
the forest to the edge of the lake. Three trees,
squared, and laid side by side, form the bottom of
the trough ; the tree in the middle having its surface hollowed, so that a rill of water received from
distance to distance, over the side of the trough,
may be conveyed along the bottom, and preserve it
moist. Adjoining to the central part, (of the trough,)
other trees, also squared, are laid parallel to the
former, in such a manner as to form a trough,
rounded in the interior, and of such dimensions as
to allow the largest trees to lie, or to move along
quite readily. When the direction of the trough
turns, or has any bending, of which there are many,
its sides are made higher and stronger, especially on
the convex side, or that from which it bends, so as
to provide against the trees bolting or flying out,
which they sometimes do, in spite of every precaution. In general, the trough is from five to six feet
wide at top, and from three to four in depth, vary-
ing, however, in different places, according to circumstances.

" This singular road has been constructed at considerable expence ; though, as it goes, almost for its
whole length, through a forest, the materials of construction were at hand, and of small value. It contains, we were told, thirty thousand trees ; it is, in
general, supported on cross timbers, that are them-

selves supported by uprights fixed in the ground;
and these cross timbers are sometimes close to the
surface; they are occasionally under it, and sometimes elevated to a great height above it. It crosses
in its way three great ravines, one at the height of
64 feet, another at the height of 103, and the third,
where it goes along the face of a rock, at that of
157; in two places it is conveyed under ground. It
was finished in 1812.

" The trees which descend by this conveyance are
spruce firs, very straight, and of great size. All
their branches are lopped off; they are stripped of
the bark, and the surface, of course, made tolerably
smooth. The trees, or logs, of which the trough is
built, are dressed with the axe, but without much
care.

" All being thus prepared, the tree is launched
with the root end foremost, into the steep part of the
trough, and in a few seconds acquires such a velocity as enables it to reach the lake in the short space
of six minutes; a result altogether astonishing,
when it is considered that the distance is more than
eight miles, that the average declivity is but one foot
in seventeen, and that the route which the trees
have to follow is often circuitous, and in some places
almost horizontal.

" Where large bodies are moved with such velocity
as has now been described, and so tremendous a force
of course produced, every thing had need to be done

with the utmost regularity; every obstacle carefully
removed that can obstruct the motion, or that might
suffer by so fearful a collision. Every thing, accordingly, with regard to launching off the trees, is directed by telegraphic signals. All along the slide,
men are stationed, at different distances, from half
a mile to three quarters, or more, but so that every
station may be seen from the next, both above and
below. At each of these stations, also, is a telegraph,
consisting of a large board like a door, that turns at
its middle on a horizontal axle. When the board is
placed upright, it is seen from the two adjacent stations; when it is turned horizontally, or rather parallel to the surface of the ground, it is invisible from
both. When the tree is launched from the top, a
signal is made, by turning the board upright; the
same is followed by the rest, and thus the information is conveyed, almost instantaneously, all along
the slide, that a tree is now on its way. Bye and
bye, to any one that is stationed on the side, even to
those at a great distance, the same is announced by
the roaring of the tree itself, which becomes always
louder and louder; the tree comes in sight when it
is perhaps half a mile distant, and in an instant after
shoots past, with the noise of thunder and the rapidity of lightning. As soon as it has reached the
bottom, the lowest telegraph is turned down, the
signal passes along all the stations, and the workmen
at the top are informed that the tree has arrived in

safety. Another is set off as expeditiously as possible ; the moment is announced, as before, and the
same process is repeated, till all the trees that have
been got in readiness for that day have been sent
down into the lake.

" When a tree sticks by accident, or when it flies
out, a signal is made from the nearest station, by
half depressing the board, and the workmen from
above and below come to assist in getting out the
tree that has stuck, or correcting any thing that is
wrong in the slide, from the springing of a beam in
the slide ; and thus the interruption to the work is
rendered as short as possible.

" We saw five trees come down ; the place where
we stood was near the lower end, and the declivity
was inconsiderable, (the bottom of the slide nearly
resting on the surface,) yet the trees passed with astonishing rapidity. The greatest of them was a spruce
fir a hundred feet long, four feet in diameter at the
lower end, and one foot at the upper. The greatest trees are those that descend with the greatest rapidity ; and the velocity as well as the roaring of
this one was evidently greater than of the rest. A
tree must be very large, to descend at all in this
manner; a tree, Mr Rupp informed us, that was
only half the dimensions of the preceding, and
therefore only an eighth part of its weight, would
not be able to make its way from the top to the bottom. One of the trees that we saw broke by some

accident into two ; the lighter part stopped almost immediately, and the remaining part came to rest soon after. This is a valuable fact ; it appears from it that the friction is not in proportion to the weight, but becomes relatively less as the weight increases, contrary to the opinion that is generally received.

" In viewing the descent of the trees, my nephew and I stood quite close to the edge of the trough, not being more interested about anything than to experience the impression which the near view of so singular an object must make on a spectator. The noise, the rapidity of the motion, the magnitude of the moving body, and the force with which it seemed to shake the trough as it passed, were altogether very formidable, and conveyed an idea of danger much greater than the reality. Our guide refused to partake of our amusement ; he retreated behind a tree at some distance, where he had the consolation to be assured by Mr Rupp, that he was no safer than we were, as a tree, when it happened to bolt from the trough, would often cut the standing trees clear over. During the whole time the slide has existed, there have been three or four fatal accidents, and one instance was the consequence of excessive temerity.

" I have mentioned that a provision was made for keeping the bottom of the trough wet ; this is a very useful precaution ; the friction is greatly diminished, and the swiftness is greatly increased by that means.

In rainy weather the trees move much faster than in
dry. We were assured that when the trough was
every where in its most perfect condition, the weather
wet, and the trees very large, the descent was sometimes made in as short a time as three minutes.

" The trees thus brought down into the Lake of
Lucerne are formed into rafts, and floated down the
very rapid stream of the Reuss, by which the lake
discharges its waters first into the Aar, and then
into the Rhine. By this conveyance, which is all
of it in streams of great rapidity, the trees sometimes
reach Basle, in a few days after they have left Lucerne; and there the immediate concern of the
Alpnach company terminated. They still continue
to be navigated down the Rhine in rafts to Holland,
and are afloat in the German Ocean in less than a
month from having descended from the side of Pilatus, a very inland mountain, not less than a
thousand miles distant. The late Emperor of France
had made a contract for all the timber thus brought
down.

" From the phenomena just described, I have deduced several conclusions, of which at present I can
only give a very general account, without entering
into any of the mathematical reasonings on which
they rest.

" 1. The rapidity of the descent is so extraordinary, it is so much greater than any thing that could
have been anticipated, exceeding that of a horse at

full speed, nearly in the ratio of 3 to 2, that the account seems to tread on the very verge of possibility, and to touch the line that divides between what
may, and what cannot exist. The same question,
therefore, I have no doubt, has occurred to many
that occurred to myself, when I first heard of this
extraordinary phenomenon.

 " Is it possible that even if there were no friction,
and if a body was accelerated along the line of swiftest descent, from a point 2500 feet above another,
and horizontally distant from it by 44,000, that it
could arrive at that lower point in three or even in
six minutes? This was the first question that occurred to me, and at a distance from books as I was
then, and in no condition to undertake any nice or
difficult calculation, I could only satisfy myself by
a rude approximation, that there was nothing in the
reported circumstance that was without the limits of
possibility. Had the result of the calculation been
contrary, I should not only have disbelieved the report, but I should have doubted the testimony of my
own senses.

 " From a more accurate calculation I find that if no
friction nor resistance took place, and if the moving
body was allowed to take its flight in the line of the
swiftest descent, that it would do so in less than sixty-
six seconds. This is the minimum then of time, and
we may rest assured, while the laws of nature continue the same that they are now, that no body, in

the circumstances just described, can perform its journey in less time than the above.

" But though the descent of the trees at Alpnach contains nothing inconsistent with the acceleration of bodies by gravity, it is not to be reconciled with the notions concerning friction, that are usually received even in the scientific world.

" It is common to consider friction as a force bearing a certain proportion to the weight of the body moved, and as retarding the body by a force proportional to its weight, amounting to a fourth or fifth part, or when least to a tenth or twelfth part of gravity. A body, therefore, that was descending along an inclined plane, would be accelerated by its own gravity, minus the force of friction, a constant force that increased in proportion to the body.

" Now, in the present case, it will soon appear that the retardation is vastly less than would arise from any of these suppositions.

" Supposing it to be true, that friction in a given instance (the surface, the inclination, and the weight, being all given) acts as a uniformly retarding force, I have found that a body sliding along an inclined surface, under the acceleration of gravity, and the retardation of friction, will be accelerated, so that it will have at every point the velocity that would be acquired by falling by its own gravity from a line inclined to the horizon, that is drawn from the point where the body began to move, and that makes with the horizon an angle, the tangent of which is the fraction, that denotes the ratio of friction to gravity. The velocity of the moving body is therefore as the square root, of the portion of a vertical passing through the body, and reaching up to the line just mentioned, or the line of no acceleration.

" As the trees at Alpnach enter the lake with a considerable velocity, it is evident that the line of no acceleration, drawn from the top of the slide, does not reach the ground at the point where the slide ends, but is then still considerably above the surface ; the tangent, therefore, of the angle which that line makes with the horizon, is much less than $\frac{1}{17}$. There is reason to think that it does not in reality amount to $\frac{1}{3}$ of this, and is therefore less than $\frac{1}{50}$. It follows, then, that the friction that trees suffer in the slide is less than one-fiftieth of their weight.

" Now, from what can we suppose the small proportion that friction, in this instance, bears to the weight, to arise ? It is not that the surfaces have a great smoothness or a fine polish. The logs that form the trough are coarsely dressed with the adze, and I observed that there was not even the precaution taken of making the grain of the wood lie downward, or toward the declivity. It was so in the tree, but not in the trees which composed the slide. It is not that any lubricating substance, oil, grease,

soap, or black-lead, is interposed between their surfaces. Water is the only substance of this kind that
is applied. We have fir rubbing on fir, which is
supposed a case remarkably unfavourable to the diminution of friction. It can only arise, therefore,
from a principle that some mechanical writers have
suspected to exist, but which was never before, I
think, proved by the direct evidence of facts, namely, that the force of friction does not increase in the
proportion of the weight of the rubbing body, so
that heavy bodies are, in reality, less retarded in their
motion on an inclined surface than lighter bodies.
This the whole of the phenomena I have been describing, tend to prove, especially the fact I mentioned, that heavy trees made their way more easily
than light ones, and that a tree must be of a certain magnitude to make its way to the bottom.
Friction, therefore, does not bear even in the same
materials a given ratio to the weight, but a ratio
that evidently decreases as the weight increases ; so
that, in a fir of ordinary size it is $\frac{1}{12}$, or $\frac{1}{20}$, in one
of 100 feet in length it is between $\frac{1}{50}$ and $\frac{1}{60}$. According to what law this change takes place, it
would be most useful to investigate ; it is an inquiry for those engineers who have strong machinery
and great power ready at command.

" I must observe also, that I strongly suspect that
friction diminishes with the velocity of the moving,
or sliding body. That it passes all at once when a

body begins to move, to be only half of what it was when the body was at rest, is quite certain, and is proved by many experiments. It seems to me not unlikely that the same progress continues as the motion becomes greater. Perhaps in as much as friction is concerned, the pressure is lessened by the velocity, and the poet was not so far mistaken as he is generally supposed to be, when he said of his heroine,

> Illa vel intactae segetis per summa volaret
> Gramina, nec teneras cursu laesisset aristas.

However that be, we have a strong example here of the danger of concluding in many of the researches of mechanics, from experiments made on a small scale to the practice that is to be proceeded on in a great one. It requires some attention to enable us to discriminate between the cases where we can safely proceed from the small to the great, and those in which we cannot. A man, from finding that bodies of a pound or half a pound are in equilibrio when their distances from the fulcrum are inversely as their weights, might, without danger of error, transfer the conclusion to weights of hundreds of tons, or to whole planets, were it possible to make the experiment on so large a scale. But when he finds that the friction of a body of a pound, or a hundred weight, is one-fourth of the weight, he cannot, with equal safety, presume that the same will hold when bodies of immense weight and size come to rub against one another. There are many other cases of the same kind. In general, when our experiments lead to the knowledge of a fact and not of a principle, there is caution required in extending the conclusions beyond the limits by which the experiments have been confined. This is the case with the experiments on friction, where we know only facts, and have no principle to guide us ; that is, we have not been able to connect the facts with any of the known and measurable properties of body. In the case of the lever, we have connected the fact with the inertia of matter, and the equality of action and reaction. We have, therefore, a right to repose confidence on the one, when extended, though not on the other.

" That friction belongs to the cases in which great caution is necessary in extending the conclusions of experiments, is indeed most strongly evinced by the operations that have now been described, the result of which is such as could not have been anticipated from those experiments. The danger here, however, is quite of an opposite kind from that which commonly takes place in such instances. The experiments on the small scale, usually represent the thing as more easy than it is upon the great, and engage us in attempts that prove abortive, and are followed by disappointments and even ruin. In the present case, the experiments on the small scale re-

present the thing as more difficult than when tried
on a great one it is found to be, and would lead us,
by an error, the direct opposite of the last, to conclude things to be impracticable that may be carried
into effect with ease. Had the ingenious inventor
of the slide at Alpnach been better acquainted with
the received theories of friction, or the experiments
on which they are founded, even those that are the
best, and on the greatest scale, such as those of another most skilful engineer, M. Coulomb, or had he
placed more faith in them, he would never have attempted the great work in which he has so eminently succeeded."

No. III.

Kinneil House, Jan. 9, 1822.

My Dear Sir,

I am sorry that you have not been able to
send me the Biographical Account of your Uncle with
the last alterations and corrections. I can, therefore, only say, that, in the state in which I saw it, I
read it with entire satisfaction, and I have no doubt
that it has since been improved in consequence of
the suggestions of your other friends.

Considering the truly filial relation in which you stood to your uncle from the period of your childhood, I think you have judged wisely in abstaining from any attempt to appreciate his scientific and literary merits. Indeed, in my opinion, such an attempt would have been wholly out of place, in front of a publication exhibiting a combination of the soundest philosophy, and of the profoundest science, with powers of eloquence and skill in composition which place the Author in the first rank of our classic writers. As to those features of his character which are less known to the public, a faithful and perfect resemblance is preserved in the masterly portrait of Mr Jeffrey, which you will, no doubt, add to your own Memoir.

Had the state of my health permitted, I should have had much satisfaction in offering the best tribute in my power to the memory of my excellent and illustrious friend ; but my late indisposition still deprives me in a great measure of the use of my right hand, and it is not without difficulty that I have been able to dictate these few sentences.

I am, my Dear Sir,

Yours most truly,

DUGALD STEWART.

To Dr James Playfair.

TABLE

OF

CONTENTS.

———

3. Position of the Strata. Page 56

SECTION II.

PHENOMENA PECULIAR TO UNSTRATIFIED BODIES.

1. Metallic Veins. p. 72

2. Whinstone. Page 81

3. Granite. p. 95

NOTES AND ADDITIONS.

Saussure between Nice and Genoa, *ib.* Remarks on it,
§ 186.

NOTE XII.—Elevation and Inflection of the Strata.
Page 216

NOTE XIII.—Metallic Veins.   p. 244

Note xiv.—On Whinstone. Page 264

Mount, Cornwall, § 285. Fragments of schistus contained in granite, § 287.

about the tides, § 380, 381,—and about the formation of sand banks, § 384.

NOTE XX.—Inequalities of the Planetary Motions.   Page 428

These inequalities all periodical, § 385. Circumstances on which this depends, § 386. Affinity of this conclusion to that which Dr Hutton has established with respect to the changes at the surface of the earth, § 387.

NOTE XXI.—Changes in the Apparent Level of the Sea.   p. 432

Relative level of the sea and land subject to change, § 388. Proofs that it has sunk, on the shores of this island, § 389, —on the coasts of France and Flanders, § 390, 391,—on the shores of the Baltic, § 392. This has not arisen from the depression of the sea, but from the elevation of the land, § 393, 394. The surface of the Hadriatic higher now than formerly, § 395, 396. Also of the Mediterranean, § 398. Irregularities in these changes, § 399, 400. Hypothesis of Frisi, that towards the equator the sea is every where rising, § 401. Disproved, *ib.* Conclusion, § 402.

NOTE XXII.—Fossil Bones.   p. 448

Vegetable and animal remains contained in the fossil kingdom, § 403. Of those that are enveloped or penetrated

ADVERTISEMENT.

THE Treatise here offered to the Public, was drawn up with a view of explaining Dr Hutton's Theory of the Earth in a manner more popular and perspicuous than is done in his own writings. The obscurity of these has been often complained of; and thence, no doubt, it has arisen, that so little attention has been paid to the ingenious and original speculations which they contain.

The simplest way of accomplishing the object proposed, seemed to be, to present a General Outline of the System, in one continued Discourse ; and to introduce afterwards, in the form of Notes, what farther elucidation any particular subject was thought to demand. Through the whole, I have aimed at little more than a clear exposition of facts, and a plain deduction of the conclusions grounded on them ; nor shall I claim any merit to myself, if, in the order which I have found it necessary to adopt, some arguments may have taken a new form, and some additions may have been made to a system naturally rich in the number and variety of its illustrations.

Of the qualifications which this undertaking requires, there is one that I may safely suppose myself to possess. Having been instructed by Dr
Hutton himself in his theory of the earth ; having
lived in intimate friendship with that excellent
man for several years, and almost in the daily habit
of discussing the questions here treated of ; I have
had the best opportunity of understanding his
views, and becoming acquainted with his peculiarities, whether of expression or of thought. In the
other qualifications necessary for the illustration of
a system so extensive and various, I am abundantly sensible of my deficiency, and shall therefore,
with great deference, and considerable anxiety,
wait that decision from which there is no appeal.

EDINBURGH COLLEGE, 1st *March* 1802.

ILLUSTRATIONS

HUTTONIAN THEORY.

A VERY little attention to the phenomena of the
mineral kingdom, is sufficient to convince us, that
the condition of the earth's surface has not been
the same at all times that it is at the present moment. When we observe the impressions of plants
in the heart of the hardest rocks ; when we discover trees converted into flint, and entire beds of
limestone or of marble composed of shells and corals ; we see the same individual in two states, the
most widely different from one another ; and, in
the latter instance, have a clear proof, that the present land was once deep immersed under the waters of the ocean. If to this we add, that many
masses of rock, the most solid and compact, consist of no other materials but sand and gravel ;
that, on the other hand, loose gravel, such as is
formed only in beds of rivers, or on the sea shore,

now abounds in places remote from both : if we reflect, at the same time, on the irregular and broken
figure of our continents, and the identity of the
mineral strata on opposite sides of the same valley,
or the same inlet of the sea ; we shall see abundant
reason to conclude, that the earth has been the
theatre of many great revolutions, and that nothing on its surface has been exempted from their
effects.

To trace the series of these revolutions, to explain their causes, and thus to connect together all
the indications of change that are found in the mineral kingdom, is the proper object of a THEORY
OF THE EARTH.

But, though the attention of men may be turned to the theory of the earth by a very superficial
acquaintance with the phenomena of geology, the
formation of such a theory requires an accurate
and extensive examination of those phenomena,
and is inconsistent with any but a very advanced
state of the physical sciences.   There is, perhaps,
in those sciences, no research more arduous than
this ; none certainly where the subject is so complex ; where the appearances are so extremely diversified, or so widely scattered, and where the
causes that have operated are so remote from the
sphere of ordinary observation.   Hence the attempts
to form a theory of the earth are of very modern
origin, and as, from the simplicity of its subject,

astronomy is the eldest, so, on account of the com-
plexness of its subject, geology is the youngest of
the sciences.

It is foreign from the present purpose to enter
on any history of the systems that, since the rise
of this branch of science, have been invented to
explain the phenomena of the mineral kingdom.
It is sufficient to remark, that these systems are
usually reduced to two classes, according as they refer the origin of terrestrial bodies to FIRE or to WA-
TER; and that, conformably to this division, their
followers have of late been distinguished by the
fanciful names of *Vulcanists* and *Neptunists*. To
the former of these Dr HUTTON belongs much
more than to the latter; though, as he employs
the agency both of fire and of water in his system,
he cannot, in strict propriety, be arranged with
either.

In the succinct account which I am now about
to give of this system, I shall consider the mineral
kingdom as divided into two parts, namely, stratified and unstratified substances. I shall treat,
first, of the phenomena peculiar to the stratified;
next, of those peculiar to the unstratified; and,
lastly, of the phenomena common to both. Beginning, then, with the first, the subject naturally
divides itself into three branches; viz. the *materials*, the *consolidation*, and the *position* of the
strata.

SECTION I.

OF THE PHENOMENA PECULIAR TO STRATIFIED
BODIES.

1. *Materials of the Strata.*

1. It is well known that, on removing the loose
earth which forms the immediate surface of the
land, we come to the solid rock, of which a great
proportion is found to be regularly disposed in
strata, or beds of determinate thickness, inclined at
different angles to the horizon, but separated from
one another by equidistant superficies, that often
maintain their parallelism to a great extent. These
strata bear such evident marks of being deposited
by water, that they are universally acknowledged
to have had their origin at the bottom of the sea ;
and it is also admitted, that the materials which
they consist of, were then either soft, or in such a
state of comminution and separation, as rendered
them capable of arrangement by the action of the
water in which they were immersed. Thus far
most of the theories of the earth agree ; but from
this point they begin to diverge, and each to assume a character and direction peculiar to itself

Dr Hutton's does so, by laying down this fundamental proposition, That in all the strata we discover proofs of the materials having existed as elements of bodies, which must have been destroyed
before the formation of those of which these materials now actually make a part. *

2. The calcareous strata are the portion of the
mineral kingdom that gives the clearest testimony
to the truth of this assertion. They often contain
shells, corals, and other exuviæ of marine animals
in so great abundance, that they appear to be composed of no other materials. Though these remains of organized bodies are now converted into
stone or into spar, their shape and interior structure are often so well preserved, that the species of
animal or plant of which they once made a part,
can still be distinguished and pointed out among
the living inhabitants of the ocean.

Others of the calcareous strata appear to be composed of fragments of some ancient rocks, which,
after having been broken, have been again united
into a compact stone. In these we find pieces
clearly marked as having been once continuous
but now placed at a distance from one another, and
exhibiting exactly the same appearances as if they
floated in a fluid of the same specific gravity with
themselves.

---

* Hutton's Theory, Vol. I. p. 20, &c.

From these, therefore, and a variety of similar
appearances, Dr Hutton concludes, that the materials of all the calcareous strata have been furnished, either from the dissolution of former strata,
or from the remains of organized bodies.   But,
though this conclusion is meant to be extended to
all the calcareous strata, it is not asserted that every
cubic inch of marble or of limestone contains in it
the characters of its former condition, and of the
changes through which it has passed.   It may,
however, be safely affirmed, that there is scarce any
entire stratum where such characters are not to be
found.   These must be held as decisive with respect
to the whole system of strata to which they belong;
they prove the existence of calcareous rocks before
the formation of the present ; and, as the destruction of those is evidently adequate to the supply of
the materials of these that we now see, to look for
any other supply were superfluous, and could only
embarrass our reasonings by the introduction of
unnecessary hypotheses. *

3. The same conclusions result from an examination of the siliceous strata ; under which we may
comprehend the common sandstone, and also those
pudding-stones or breccias where the gravel consists of quartz.   In all these instances, it is plain,
that the sand or gravel existed in a state quite loose

---

* NOTE I.

and unconnected, at the bottom of the sea, previous
to its consolidation into stone. But such bodies of
gravel or sand could only be formed from the attrition of large masses of quartz, or from the dissolution of such sandstone strata as exist at present;
for it will hardly be alleged, that sand is a crystallization of quartz, formed from that substance, when
it passes from a fluid to a solid state.

Those pudding-stones in which the gravel is
round and polished, carry the conclusion still farther, as such gravel can only be formed in the beds
of rivers or on the shores of the sea; for, in the
depths of the ocean, though currents are known to
exist, yet there can be no motion of the water sufficiently rapid to produce the attrition required to
give a round figure and smooth surface to hard
and irregular pieces of stone. There must have
existed, therefore, not only a sea, but continents,
previously to the formation of the present strata.

The same thing is clearly shown by those petrifactions of wood, where, though the vegetable structure is perfectly preserved, the whole mass is siliceous, and has, perhaps, been found in the heart of
some mountain, deep imbedded in the solid rock.

4. Characters of the same import are also found
among the argillaceous strata, though perhaps more
rarely than among the calcareous or siliceous. Such
are the impressions of the leaves and stems of vegetables; also the bodies of fish and amphibious

animals, found very often in the different kinds of argillaceous schistus, and in most instances having the figure accurately preserved, but the substance of the animal replaced by clay or pyrites. These are all remains of ancient seas or continents ; the latter of which have long since disappeared from the surface of the earth, but have still their memory preserved in those archives, where nature has recorded the revolutions of the globe.

5. Among bituminous bodies, pit-coal is the only one which constitutes regular and extensive strata ; and no fossil has its origin from the waste of former continents, marked by stronger and more distinct characters. Not to mention that the coal strata are alternated with those that have been already enumerated, and that they often contain shells and corals, perfectly mineralized, it is sufficient to remark, that there are entire beds of this fossil, which appear to consist wholly of wood, and in which the fibrous structure is perfectly preserved. From these instances, the appearances of vegetable structure may be traced through all possible gradations, down to an evanescent state. This last state is undoubtedly the most common ; and though coal does not then, on bare inspection, make known its vegetable origin, yet, if we take it in connection with the other terms of the series, as we may call them ; if we consider that the two extremes, viz. coal, with the vegetable structure per-

fect, and coal without any such structure visible,
are often found in the same or in contiguous beds ;
and, if we remark, that through all these gradations coal contains nearly the same chemical elements, and yields, on analysis, bitumen and charcoal, combined with a greater or less proportion of
earth : if we take all these circumstances into account, we cannot doubt that this fossil is every
where the same, and derives its origin from the
trees and plants that grew on the surface of the
earth before the formation of the present land.

6. Dr Hutton has further observed, that if
those ancient continents were at all similar to the
present, we can be at no loss to account for the
want of any distinct mark of vegetable organization in the greater part of the coal strata. It is
plain, that the daily waste of animal and vegetable
substances on the surface of the earth, must disen-
gage a great quantity of oily as well as carbonic
matter, which, with whatever element it is at first
combined, is ultimately delivered into the ocean.
Thus, the oily or fuliginous parts of animal and
vegetable substances, let loose by burning, first
ascend into the atmosphere, but are at length precipitated, and either fall immediately into the sea,
or are, in part at least, washed down into it from
the land. From other causes also, much vegetable
matter is carried down by the rivers ; and the whole
quantity of animal and vegetable substances thus

delivered into the sea, must be very considerable, amounting annually to the whole residuum of those substances, not employed in the maintenance or re· production of animal and vegetable bodies. Whether chemically united to the waters of the ocean, or simply suspended in them, this matter is at last precipitated, and, mingling with earthy substances, is formed into strata, the place of which will be determined by the currents, the position of the present continents, and many other circumstances not easily enumerated.

If, then, an order of things similar to what we now see, existed before the formation of the present strata, it would necessarily happen, that the animal and vegetable substances, diffused through the ocean, being separated from the water, would be deposited at the bottom of the sea, and, in the course of ages, would form beds, less or more pure, according to the quantity of earth and other substances deposited at the same time. These beds being consolidated and mineralized by operations that are afterwards to be considered, have been converted into pit-coal, the parts of which are impalpable, and retain nothing of their primitive structure. *

If, then, the formation of coal from animal and vegetable bodies be admitted, the general position

---

* Note ii.
12

which derives the origin of the strata from the
waste of former land, as it is applicable to all the
kinds already enumerated, and of course to all
those with which they are alternated, comprehends
a very large portion of the earth's surface. It
comprehends, indeed, all the strata usually distinguished by the name of *Secondary ;* but there is
another great division of the mineral kingdom, viz.
the rocks, called *Primitive,* which, as they are never alternated with the secondary, but are always
inferior to them, must be further examined, before
we can decide whether the same conclusion extends
to them or not.

7. Here it must be carefully observed, that,
among the primary rocks, the granite is not meant
to be included, except where that stone is stratified,
and either coincides with veined granite or with
gneiss. The primitive strata, in Dr Hutton's
theory, comprehend, besides gneiss, the micaceous,
chlorite, hornblende, and siliceous schistus, together
with slate, and some other kinds of argillite ; to
which we must add, serpentine, micaceous limestone, and the greater part of marbles. These are
mostly distinguished by their laminated structure,
by having their planes much elevated with respect
to the horizon, and by belonging more to the mountainous than the level parts of the earth's surface.
They rarely contain vestiges of organized bodies ;
so rarely, indeed, that they were called primitive

by the geologists who first distinguished them from
other rocks, on the supposition of their being part
of the primeval nucleus of the globe, which had
never undergone any change whatsoever ; but this,
I believe, has now almost ceased to be the opinion
of any geologist. *   The Neptunists hold the rocks,
here enumerated, and also granite, to be produced
by aqueous deposition ; but maintain them to be in
the strictest sense primeval, and of a formation antecedent to all organized bodies.

8. In opposition to this, Dr Hutton maintained,
that the primary schistus, like all the other strata,
was formed of materials deposited at the bottom of
the sea, and collected from the waste of rocks still
more ancient.   When, therefore, he conformed to
the received language of mineralogists, by calling
these strata primitive, he only meant to describe
them as more ancient than any other strata now existing, but not as more ancient than any that ever had
existed.   They are distinguished, in his system, by
the name of *Primary*, rather than of *Primitive*
strata.

That the account now given of their origin is
well founded, may be proved by unquestionable
facts.   For, first, though, agreeably to the observation just made, the ancient strata do but rarely
contain any remains of organized bodies, they are

----

* NOTE III.
4

not entirely destitute of them. Different places in this island have been pointed out by Dr Hutton, where marine objects have been discovered in primary limestone, either by himself or others, and it would not be difficult to add more instances of the same kind. * In Dauphiné, coal, which is certainly a derivative substance, has been found among mountains which have a title to the character of primitive, such as no one will dispute. These facts put the composition of such rocks from loose materials, beyond all doubt, and also prove their formation to be posterior to the existence of an animal and vegetable system. They do indeed prove this in the strictest sense, only of the particular beds in which they are found; but as these beds are in all other respects as much to be accounted primary as any part of the mineral kingdom, it is evident that the negative instances are here of no force, and that nothing can be gained to the adversaries of this opinion by denying it in general, if they are obliged to admit it in a single case.

9. Again, it is certain, as Dr Hutton remarks, that there are few considerable bodies of schistus, even the most decidedly primitive, where sand and gravel may not in some parts be observed. Indeed, it is not only true that they are to be found in some parts of them; but, in fact, among many of

---

* NOTE IV.

the primitive mountains, we find large tracts, composed entirely of a schistose and much indurated
sandstone, in beds highly inclined, sometimes
alone, sometimes alternated with other schisti. In
many of them, the sand of which they consist appears to be entirely of granite, from the detritus of
which rock it should seem that they were chiefly
formed.

10. Thus we conclude, that the strata both primary and secondary, both those of ancient and
those of more recent origin, have had their materials furnished from the ruins of former continents,
from the dissolution of rocks, or the destruction of
animal or vegetable bodies, similar, at least in some
respects, to those that now occupy the surface of
the earth. This conclusion is not indeed proved of
every individual portion of rock, but it is demonstrated of many and large parts, and those scattered
indifferently through all the varieties of the strata;
and therefore, from the rules of the strictest reasoning, we must infer, that the whole is derived
from the same origin. *

Thus far concerning the materials of the strata;
and, as these were originally loose and unconnected, we must next consider by what means they
were consolidated into stone.

---

* NOTE V.

2. *Consolidation of the Strata.*

11. Though Dr Hutton has no where defined the meaning of the term consolidation, he has been scrupulously exact in using it constantly in the same sense. He understands by it, not merely that quality in a hard body, by which its parts co-here together, but also that by which it fills up the space comprehended within its surface, being to sense without porosity, and impervious to air and moisture.

Now, a porous mass of unconnected materials, such as the strata appear originally to have been, can acquire hardness and solidity only in two ways, that is, either when it is first reduced by heat into a state of fusion, or at least of softness, and after-wards permitted to cool ; or when matter that is dissolved in some fluid menstruum, is introduced along with that menstruum into the porous mass, and, being deposited, forms a cement by which the whole is rendered firm and compact. Fire and water, therefore, are the only two physical agents to which we can ascribe the consolidation of the strata ; and, in order to determine to which of them that effect is to be attributed, we must inquire whether there are any certain characters that dis-tinguish the action of the one from that of the other, and which may be compared with the phe-

nomena actually observed among mineral substances.

12. First, then, it is evident, that the consolidation produced by the action of water, or of any
other fluid menstruum, in the manner just referred
to, must necessarily be imperfect, and can never
entirely banish the porosity of the mass.   For the
bulk of the solvent, and of the matter it contained
in solution, being greater than the bulk of either
taken singly, when the latter was deposited, the
former would have sufficient room left, and would
continue to occupy a certain space in the interior
of the strata.   A liquid solvent, therefore, could
never shut up the pores of a body to the entire exclusion of itself; and, had mineral substances been
consolidated, as here supposed, the solvent ought
either to remain within them in a liquid state, or,
if evaporated, should have left the pores empty,
and the body pervious to water.   Neither of these,
however, is the fact ; many stratified bodies are
perfectly impervious to water, and few mineral
substances contain water in a liquid state.   That
they sometimes contain it, chemically united to
them, is no proof of their solidity having been
brought about by that fluid ; for such chemical
union is as consistent with the supposition of igneous as of aqueous consolidation, since the region in
which the fire was applied, on every hypothesis,
must have abounded with humidity.

13. Again, if water was the solvent by which the consolidating matter was introduced into the interstices of the strata, that matter could consist only of such substances as are soluble in water, whereas it consists of a vast variety of substances, altogether insoluble either in it, or in any single menstruum whatsoever. The strata are consolidated, for example, by quartz, by fluor, by feldspar, and by all the metals, in their endless combinations with sulphureous bodies. To affirm that water was ever capable of dissolving these substances, is to ascribe to it powers which it confessedly has not at present ; and, therefore, it is to introduce an hypothesis, not merely gratuitous, but one which, physically speaking, is absurd and impossible.

This is not all, however ; for, even if this difficulty were to be passed over, it would still be required to explain, how the water, which, together with the matter which it held in solution, had insinuated itself into the pores of the strata, became suddenly disposed to deposit that matter, and to allow it, by crystallization or concretion, to assume a solid form. * The Neptunists must either assign a sufficient reason for this great and universal change, or must expect to see their system treated as an inartificial accumulation of hypotheses which assigns opposite virtues to the same subject, and is

* Note vi.

alike at variance with nature and with itself; in a word, a system that might pass for the invention of an age, when as yet sound philosophy had not alighted on the earth, nor taught man that he is but the minister and interpreter of nature, and can neither extend his power nor his knowledge a hair's-breadth beyond his experience and observation of the present order of things. *

14. Such are the more obvious, but I think un-answerable objections, that may be urged against the aqueous consolidation of the strata. It is true, that stony concretions, some of them much indu-rated, are formed in the humid way under our eyes. Very particular conditions, however, are re-quired for that purpose, and conditions such as can hardly have existed at the bottom of the sea. First, The water must dissolve the substance of which the concretion is to be formed, as it actually does in the case of calcareous, and in certain cir-cumstances, in that of siliceous, earth. Secondly, It must be separated from that substance, as by evaporation, or by a combination of the matter dis-solved with some third substance, to which it has a greater affinity than to water, so as to form with it

---

* Homo naturæ minister, et interpres tantùm facit et in-telligit, quantùm de naturæ ordine re, vel mente, observa-verit: nec amplius scit, aut potest.

Nov. Org. Lib. i. Aph. 1.

an insoluble compound. Lastly, The water that is deprived of its solution must be carried off, and more of that which contains the solution must be supplied, as sometimes happens where water runs in a stream, or drops from the roof of a cavern. The two last conditions are peculiarly inapplicable to the bottom of the sea, where the state of the surrounding fluid would neither permit the water that was deprived of its solution from being drawn off, nor that which contained the solution from succeeding it.

It is further to be observed, that the consolidation of stalactitical concretions, that is, the filling up of their pores, is always imperfect, and is brought about by the repeated action of the fluid running through the porous mass, and continuing to deposit there some of the matter it holds in solution. This, which is properly infiltration, is incompatible with the nature of a fluid, either nearly, or altogether quiescent.

15. In order to judge whether objections of equal weight can be opposed to the hypothesis of igneous consolidation, we must attend to a very important remark, first made by Dr Hutton, and applied with wonderful success to explain the most mysterious phenomena of the mineral kingdom.

It is certain, that the effects of fire on bodies vary with the circumstances under which it is applied to them, and, therefore, a considerable allowance

must be made, if we would compare the operation
of that element when it consolidated the strata,
with the results of our daily experience.   The materials of the strata were disposed, as we have already seen, loose and unconnected, at the bottom
of the sea; that is, even on the most moderate
estimation, at the depth of several miles under its
surface.   At this depth, and under the pressure of
a column of water of so great a height, the action
of heat would differ much from that which we observe here upon the surface ; and, though our experience does not enable us to compute with accuracy the amount of this difference, it nevertheless
points out the direction in which it must lie, and
even marks certain limits to which it would probably extend.

The tendency of an increased pressure on the
bodies to which heat is applied, is to restrain the
volatility of those parts which otherwise would
make their escape, and to force them to endure a
more intense action of heat.   At a certain depth
under the surface of the sea, the power even of a
very intense heat might therefore be unable to
drive off the oily or bituminous parts from the inflammable matter there deposited, so that, when
the heat was withdrawn, these principles might be
found still united to the earthy and carbonic parts,
forming a substance very unlike the residuum obtained after combustion under a pressure no greater

than the weight of the atmosphere. It is in like manner reasonable to believe, that, on the application of heat to calcareous bodies under great compression, the carbonic gas would be forced to remain; the generation of quicklime would be prevented, and the whole might be softened, or even completely melted; which last effect, though not directly deducible from any experiment yet made, is rendered very probable, from the analogy of certain chemical phenomena.

16. An analogy of this kind, derived from a property of the barytic earth, was suggested by that excellent chemist and philosopher, the late Dr Black. The barytic earth, as is well known, has a stronger attraction for fixed air than common calcareous earth has, so that the carbonate of barytes is able to endure a great degree of heat before its fixed air is expelled. Accordingly, when exposed to an increasing heat, at a certain temperature, it is brought into fusion, the fixed air still remaining united to it : if the heat be further increased, the air is driven off, the earth loses its fluidity, and appears in a caustic state. Here, it is plain, that the barytic earth, which is infusible, or very refractory, *per se,* as well as the calcareous, owes its fusibility to the presence of the fixed air; and it is therefore probable, that the same thing would happen to the calcareous earth, if by any means the fixed air were prevented from escaping when great heat is applied to it. This escape of the fixed air

is exactly what the compression in the subterraneous
regions is calculated to prevent, and therefore we
are not to wonder if, among the calcareous strata,
we find marks of actual fusion having taken place.*

17. These effects of pressure to resist the decomposition, and augment the fusibility of bodies,
being once supposed, we shall find little difficulty
in conceiving the consolidation of the strata by
heat, since the intervals between the loose materials
of which they originally consisted may have been
closed, either by the softening of those materials,
or by the introduction of foreign matter among
them, in the state of a fluid, or of an elastic vapour.
No objection to this hypothesis can arise from the
considerations stated in the preceding case ; the
solvent here employed would want no pores to
lodge in after its work was completed, nor would
it find any difficulty in making its retreat through
the densest and most solid substances in the mineral kingdom. Neither can its incapacity to dissolve
the bodies submitted to its action be alleged. Heat
is the most powerful and most general of all sol-
vents ; and, though some bodies, such as the calcareous, are able to resist its force on the surface
of the earth, yet, as has just been shown, it is perfectly agreeable to analogy to suppose, that, under
great pressure, the carbonic state of the lime being
preserved, the purest limestone or marble might be

---

* NOTE VII.

softened, or even melted. With respect to other substances, less doubt of their fusibility is entertained ; and though, in our experiments, the refractory nature of siliceous earth has not been completely subdued, a degree of softness and an incipient fusion have nevertheless been induced.

Thus it appears, in general, that the same difficulties do not press against the two theories of aqueous and of igneous consolidation ; and, that the latter employs an agent incomparably more powerful than the former, of more general activity, and, what is of infinite importance in a philosophical theory, vastly more definite in the laws of its operation.

18. A more particular examination of the different kinds of fossils will confirm this conclusion, and will show, that, wherever they bear marks of having been fluid, these marks are such as characterize the fluidity of fusion, and distinguish it from that which is produced by solution in a menstruum. Dr Hutton has enumerated many of these discovered in the course of that careful and accurate examination of fossils, in which he probably never was excelled by any mineralogist. It will be sufficient here to point out a few of the most remarkable examples.

19. Fossil wood, penetrated by siliceous matter, is a substance well known to mineralogists ; it is found in great abundance in various situations, and

frequently in the heart of great bodies of rock. On examination, the siliceous matter is often observed to have penetrated the wood very unequally, so that the vegetable structure remains in some places entire ; and in other places is lost in a homogeneous mass of agate or jasper.   Where this happens, it may be remarked, that the line which separates these two parts is quite sharp and distinct, altogether different from what must have taken place, had the flinty matter been introduced into the body of the wood, by any fluid in which it was dissolved, as it would then have pervaded the whole, if not uniformly, yet with a regular gradation.   In those specimens of fossil wood that are partly penetrated by agate, and partly not penetrated at all, the same sharpness of termination may be remarked, and is an appearance highly characteristic of the fluidity produced by fusion.

20. The round nodules of flint that are found in chalk, quite insulated and separate from one another, afford an argument of the same kind ; since the flinty matter, if it had been carried into the chalk by any solvent, must have been deposited with a certain degree of uniformity, and would not now appear collected into separate masses, without any trace of its existence in the intermediate parts. On the other hand, if we conceive the melted flint to have been forcibly injected among the chalk, and to have penetrated it, somewhat as mercury

8

may, by pressure, be made to penetrate through
the pores of wood, it might, on cooling, exhibit
the same appearances that the chalk-beds of England do actually present us with.

The siliceous pudding-stone is an instance closely connected with the two last ; in it we find both
the pebbles, and the cement which unites them,
consisting of flint equally hard and consolidated ;
and this circumstance, for which it is impossible to
account by infiltration, or the insinuation of an
aqueous solvent, is perfectly consistent with the
supposition, that a stream of melted flint has been
forcibly injected among a mass of loose gravel.

21. The common grit, or sandstone, though it
certainly gives no indication of having possessed
fluidity, is strongly expressive of the effects of heat.
It is so, especially in those instances where the
particles of quartzy sand, of which it is composed,
are firmly and closely united, without the help of
any cementing substance whatsoever. This appearance, which is very common, seems to be quite
inconsistent with every idea of consolidation, except an incipient fusion, which, with the assistance
of a suitable compression, has enabled the particles
of quartz to unite into stone.

It has indeed been asserted, that the mere apposition of stony particles, so as to permit their cor-
puscular attraction to take place, was sufficient to
form them into stone. To this Dr Hutton has

very well replied, that, admitting the possibility of
a hard and firm body being produced in this way,
of which, however, we have no proof, the close and
compact texture, the perfect consolidation of the
stones we are now speaking of, would still remain
to be explained, and of this it is evident that the
mere apposition of particles, and the force of their
mutual attraction, can afford no solution.

22. These proofs that the strata must have endured the action of intense heat, though immediately deduced from those of the siliceous genus
only, extend in reality to all the strata, of every
kind, with which they are found alternated.   It is
impossible that heat, of the intensity here supposed, can have acted on a particular stratum, and
not on those that are contiguous to it ; and, as
there are no strata of any kind with which the
quartzy and siliceous are not intermixed, so there
are none of which the igneous consolidation is not
thus rendered probable.   We need rest nothing,
however, on this argument, as the fossils of every
genus may be shown to speak distinctly for themselves.

23. Those of the calcareous genus do so perhaps more sparingly than the rest ; yet even
among them there are many facts, that, though taken unconnected with all others, are sufficient to
establish the action of subterraneous fire.   Such,
for example, are the calcareous breccias, composed

of fragments of marble or limestone, and not only
adapted to each other's shape, but indented into
one another, in a manner not a little resembling
the *sutures* of the human *cranium*. From such
instances, it is impossible not to infer the softness
of the calcareous fragments when they were consolidated into one mass. Now, this softness could
be induced only by heat; for it must be acknowledged, that the action of any other solvent is
quite inadequate to the softening of large fragments of stone, without dissolving them altogether.

24. In many other instances it appears certain,
that the stones of the calcareous genus have been
reduced by heat into a state of fluidity much more
perfect. Thus, the saline or finer kinds of marble, and many others that have a structure highly
crystallized, must have been softened to a degree
little short of fusion, before this crystallization
could take place. Even the petrifactions which
abound so much in limestones tend to establish the
same fact; for they possess a sparry structure, and
must have acquired that structure in their transition from a fluid to a solid state. *

25. In accounting, by the operation of heat, for
these appearances of fluidity, Dr Hutton has proceeded on the principle already laid down, as conformable to analogy, that calcareous earth, under

---

* NOTE VIII.

great compression, may have its fixed air retained
in it, notwithstanding the action of intense heat,
and may, by that means, be reduced into fusion,
or into a state approaching to it.   In all this I do
not think that he has departed from the strictest
rules of philosophical investigation.   The facts just
stated prove, that limestone was once soft, its fragments retaining at the same time their peculiar
form, an effect to which we know of none similar
but those of fire ; and, therefore, though we could
not conjecture how heat might be applied to limestone so as to melt it, instead of reducing it to a
calx, we should, nevertheless, have been forced to
suppose, that this had actually taken place in the
bowels of the earth ; and was a fact which, though
we were not able to explain it, we were not entitled
to deny.   The principle just mentioned relieves us
therefore from a difficulty, that would have embarrassed, but could not have overturned, this theory
of the earth.

26. From the arguments which the argillaceous
strata afford for the igneous consolidation of fossils,
I shall select one on which Dr Hutton used to lay
considerable stress, and which some of the adversaries of his system have endeavoured to refute.
This argument is founded on the structure of certain ironstones called *septaria*, often met with
among the argillaceous schistus, particularly in the
vicinity of coal.   These stones are usually of a

lenticular or spheroidal form, and are divided in
their interior into distinct *septa*, by veins of calcareous spar, of which one set are circular and concentric, the other rectilineal ; diverging from. the
centre of the former, and diminishing in size as
they recede from it.   Now, what is chiefly to be
remarked is, that these veins terminate before they
reach the surface of the stone ; so that the matter
with which they are filled cannot have been introduced from without by infiltration, or in any
other way whatsoever.   The only other supposition, therefore, that is left for explaining the singular structure of this fossil, is, that the whole mass
was originally fluid, and that, in cooling, the calcareous part separated from the rest, and afterwards crystallized.

27. It has been urged against this theory of the
septaria, that these stones are sometimes found
with the calcareous veins extending all the way to
the circumference, and of course communicating
with the outside.   But it must be observed, that
this fact does not affect the argument drawn from
specimens in which no such communication takes
place.   It is at best only an ambiguous instance,
that may be explained by two opposite theories, and may be reconciled either to the notion
of igneous or of aqueous consolidation : but if
there is a single close septarium in nature, it can,
of course, be explained only by one of these theo-

ries, and the other must, of necessity, be rejected.
Besides, it is plain, that a close septarium can never have been open, though an open septarium may
very well have been close ; and indeed, as this
stone is, in certain circumstances, subject to perpetual exfoliation, it would be wonderful if no one
was ever found with the calcareous veins reaching
to the surface.   With regard to the light, therefore, that they give into their own history, these
two kinds of septaria are by no means on an equal
footing ; and this may serve to show, how necessary it is, in all inductive reasoning, and particularly in a subject so complex as geology, to separate with care such phenomena as admit of two solutions, from such as admit only of one.

28. The bituminous strata come next to be considered ; and they are of great consequence in the
present argument, because their dissimilarity in so
many particulars to all other mineral substances,
renders them what Lord Bacon calls an *instantia
singularis*, having the first rank among facts subservient to inductive investigation.   But though
unlike in substance to other fossils, and composed,
as has been shown, of materials that belonged not
originally to the mineral kingdom, they agree in
many material circumstances with the strata already enumerated.   Their beds are disposed in the
same manner, and are alternated indiscriminately
with those of all the secondary rocks, and, being

formed in the same region, must have been subject
to the same accidents, and have endured the operation of the same causes. They are traversed too
like the other strata, by veins of the metals, of
spars, of basaltes, and of other substances ; and,
whatever argument may hereafter be derived from
this to prove the action of fire on the strata so traversed, is as much applicable to coal as to any other
mineral. The coal strata also contain pyrites in
great abundance, a substance that is perhaps,
more than any other, the decided progeny of fire.
This compound of metal and sulphur, which is
found in mineral bodies of every kind, I believe,
without any exception, is destroyed by the contact
of moisture, and resolved into a vitriolic salt. At
the same time it is found in the strata, not traversing them in veins, which may be supposed of more
recent formation than the strata themselves ; but
existing in the heart of the most solid rocks, often
nicely crystallized, and completely inclosed, on all
sides, without the most minute vacuity. The pyrites must have been present, therefore, when the
strata were consolidated, and it is inconceivable, if
their consolidation was brought about in the wet
way, that a substance should be so generally found
in them, the very existence of which is incompatible with humidity. This argument for the igneous
origin of the strata is applicable to them all, but

especially to those of coal, as abounding with pyrites more than any other.

29. The difficulty that here naturally presents
itself, viz. how vegetable matter, such as coal is
supposed to have been, could be exposed to the action of intense heat, without being deprived of its
inflammable part, is obviated by the principle formerly explained concerning the effects of compression.   The weight incumbent on the strata of
coal, when they were exposed to the intense heat
of the mineral regions, may have been such as to
retain the oily and bituminous, as well as sulphureous parts, though the whole was reduced almost
to fusion ; and thus, on cooling, the sulphur uniting with iron might crystallize, and assume the
form of pyrites.

30. The compression, however, has not in every
instance preserved the bituminous, in union with
the carbonic part of coal ; and hence a mark of the
operation of fire quite peculiar to this fossil, and
found in those infusible kinds of it which contain
no bitumen, and burn without flame.   These resemble, some of them very precisely, and all of
them in a great degree, the products obtained by
the distillation of the common bituminous coal ;
that is, they consist of charcoal, united to an
earthy basis in different proportions.   It is natural
therefore to conclude, that this substance was pre-

pared in the mineral regions by the action of heat,
which, in some instances, has driven off the inflammable part of the coal. That the heat should,
in some cases, have done so, is not inconsistent
with the general effect attributed to compression.
The conditions necessary for retaining the more
volatile parts, may not have been present every
where in the same degree, so that the latter,
though they could not escape, may have been forced from one part of a stratum, or body of strata,
to another.

31. In confirmation of this it must be observed,
that, as the fixed part of coal is thus found in the
bowels of the earth, separate from the volatile or
bituminous, so, in the neighbourhood of coal strata,
the latter is sometimes found without any mixture
of the former. The fountains of naphtha and petroleum are well known ; and Dr Hutton has described a stratum of limestone, lying in the centre
of a coal country, which is pervaded and tinged by
bituminous matter, through its whole mass, and has,
at the same time, many close cavities in the heart
of it, lined with calcareous spar, and containing
fossil pitch, sometimes in large pieces, sometimes in
hemispherical drops, scattered over the surface of
the cavities. This combination could only be effected by a part of the inflammable matter of the
beds of coal underneath, being driven off by heat,

and made to penetrate the limestone, while it was yet soft and pervious to heated vapours. *

32. Hitherto we have enumerated those fossils that are either not at all, or very sparingly soluble in water. There are, however, saline bodies among the mineral strata, such for instance as rock-salt, which are readily dissolved in water ; and it yet remains to examine by what cause their consolidation has been effected.

Here the theorists who consider water as the sole agent in the mineralization of fossils, are indeed delivered from one difficulty, but it is only that they may be harder pressed on by another. It cannot now be said, that the menstruum which they employ is incapable of dissolving the substances exposed to its action, as in the case of metallic or stony bodies ; but it may very well be asked, how the water came to deposit the salts which it held in solution, and to deposit them so copiously as it has done in many places, without any vestige of similar deposition in the places immediately contiguous. If they refuse to call to their assistance any other than their favourite element, they will not find it easy to answer this question, and must feel the embarrassment of a system, subject to two difficulties, so nicely, but so unhappily adjusted, that one of them is always prepared to act whenever the other is re-

---

* NOTE IX.

4

moved. If, on the other hand, they will admit the
operation of subterraneous heat, it appears possible,
that the local application of such heat may have
driven the water, in vapour, from one place to another, and by such action often repeated in the same
spot, may have produced those great accumulations
of saline matter, that are actually found in the
bowels of the earth.

33. But granting that, either in the way just
pointed out, or in some other that is unknown, the
salt and the water have been separated, some further action of heat seems requisite, before a compact, and highly indurated body, like rock-salt,
could be produced. The mere precipitation of the
salt, would, as Dr Hutton has observed, form only
an assemblage of loose crystals at the bottom of the
sea, without solidity or cohesion : and to convert
such a mass into a firm and solid rock, would require the application of such heat as was able to reduce it into fusion. The consolidation of rock-salt,
therefore, however its separation from the water is
accounted for, cannot be explained but on the hypothesis of subterraneous heat.

34. Some other phenomena that have been observed in salt mines, come in support of the same
conclusion. The salt rock of Cheshire, which lies
in thick beds, interposed between strata of an argillaceous or marly stone, and is itself mixed with a
considerable portion of the same earth, exhibits a

very great peculiarity in its structure. Though it
forms a mass extremely compact, the salt is found
to be arranged in round masses of five or six feet in
diameter, not truly spherical, but each compressed
by those that surround it, so as to have the shape of
an irregular polyhedron. These are formed of concentric coats, distinguishable from one another by
their colour, that is, probably by the greater or less
quantity of earth which they contain, so that the
roof of the mine, as it exhibits a horizontal section
of them, is divided into polygonal figures, each with
a multitude of polygons within it, having altogether no inconsiderable resemblance to a *mosaic*
pavement. In the triangular spaces without the
polygons, the salt is in coats parallel to the sides of
the polygons.

The circumstances which gave rise to this singular structure we should in vain endeavour to define ;
yet some general conclusions concerning them seem
to be within our reach. It is clear that the whole
mass of salt was fluid at once, and that the forces,
whatever they were, which gave solidity to it, and
produced the new arrangement of its particles, were
all in action at the same time. The uniformity of
the coated structure is a proof of this, and, above all,
the compression of the polyhedra, which is always
mutual, the flat side of one being turned to the flat
side of another, and never an angle to an angle,
nor an angle to a side. The coats formed as it

were round so many different centres of attraction,
is also an appearance quite inconsistent with the notion of deposition ; both these, however, are com-
patible with the notion of solidity acquired by the
refrigeration of a fluid, where the whole mass is acted on at the same time, and where no solvent remains to be disposed of after the induration of the
rest.

35. Another species of fossil salt exhibits appearances equally favourable to the theory of igneous consolidation.   This is the Trona of Africa,
which is no other than soda, or mineral alkali, in a
particular state.   The specimen of this fossil in Dr
Black's, now Dr Hope's, collection, is of a sparry
and radiated structure, and is evidently part of the
contents of a vein, having a stony crust adhering to
it, on one side, with its own sparry structure complete, on the opposite.   It contains but about one-
sixth of the water of crystallization essential to this
salt when obtained in the humid way ; and, what
is particularly to be remarked, it does not lose this
water, nor become covered with a powder, like the
common alkali, by simple exposure to the air.   It
is evident, therefore, that this fossil does not originate from mere precipitation ; and when we add,
that in its sparry structure it contains evident marks
of having once been fluid, we have little reason to
entertain much doubt concerning the principle of
its consolidation.

Thus, then, the testimony given to the operation of fire, or heat, as the consolidating power of the mineral kingdom, is not confined to a few fossils, but is general over all the strata. How far the unstratified fossils agree in supporting the same conclusion, will be afterwards examined.

3. *Position of the Strata.* *

36. We have seen of what materials the strata are composed, and by what power they have been consolidated ; we are next to inquire, from what cause it proceeds, that they are now so far removed from the region which they originally occupied, and wherefore, from being all covered by the ocean, they are at present raised in many places fifteen thousand feet above its surface. Whether this great change of relative place can be best accounted for by the depression of the sea, or the elevation of the strata themselves, remains to be considered.

Of these two suppositions, the former, at first sight, seems undoubtedly the most probable, and we feel less reluctance to suppose, that a fluid, so unstable as the ocean, has undergone the great revolution here referred to, than that the solid foun-

* Theory of the Earth, Vol. I. p. 120.

dations of the land have moved a single fathom from their place. This, however, is a mere illusion. Such a depression of the level of the sea as is here supposed, could not happen without a change proportionally great in the solid part of the globe; and, though admitted as true, will be found very inadequate to explain the present condition of the strata.

37. Supposing the appearances which clearly indicate submersion under water to reach no higher than ten thousand feet above the present level of the sea, and of course the surface of the sea to have been formerly higher by that quantity than it is now; it necessarily follows, that a bulk of water has disappeared, equal to more than a seven hundredth part of the whole magnitude of the globe. * The existence of empty caverns, of extent sufficient to contain this vast body of water, and of such a convulsion as to lay them open, and give room to the retreat of the sea, are suppositions which a philosopher could only be justified in admitting, if they promised to furnish a very complete explanation of appearances. But this justification is entirely wanting in the present case; for the retreat of the ocean to a lower level, furnishes a very partial and imperfect explanation of the phenomena of

---

* NOTE X.

geology. It will not explain the numberless remains of ancient continents that are involved, as we
have seen, in the present, unless it be supposed
that the ancient ocean, though it rose to so great a
height, had nevertheless its shores, and was the
boundary of land still higher than itself. And, as
to that which is now more immediately the object
of inquiry, the position of the strata, though the
above hypothesis would account in some sort for
the change of their place, relatively to the level of
the sea ; yet, if it shall be proved, that the strata
have changed their place relatively to each other,
and relatively to the plane cf the horizon, so as to
have had an angular motion impressed on them, it
is evident that, for these facts, the retreat of the sea
does not afford even the shadow of a theory.

·38. Now, it is certain, that many of the strata
have been moved angularly, because that, in their
original position, they must have been all nearly
horizontal.   Loose materials, such as sand and
gravel subsiding at the bottom of the sea, and having their interstices filled with water, possess a
kind of fluidity : they are disposed to yield on the
side opposite to that where the pressure is greatest,
and are therefore, in some degree, subject to the
laws of hydrostatics.   On this account they will
arrange themselves in horizontal layers ; and the
vibrations of the incumbent fluid, by impressing a

slight motion backward, and forward, on the materials of these layers, will very much assist the accuracy of their level.

It is not, however, meant to deny, that the form
of the bottom might influence, in a certain degree,
the stratification of the substances deposited on it.
The figure of the lower beds deposited on an uneven surface, would necessarily be affected by two
causes ; the inclination of that surface, on the one
hand, and the tendency to horizontality, on the
other ; but, as the former cause would grow less
powerful as the distance from the bottom increased,
the latter cause would finally prevail, so that the
upper beds would approach to horizontality, and
the lower would neither be exactly parallel to them,
nor to one another.   Whenever, therefore, we meet
with rocks, disposed in layers quite parallel to one
another, we may rest assured, that the inequalities
of the bottom have had no effect, and that no cause
has interrupted the statical tendency above explained.

Now, rocks having their layers exactly parallel,
are very common, and prove their original horizontality to have been more precise than we could ven-
ture to conclude from analogy alone.   In beds of
sandstone, for instance, nothing is more frequent
than to see the thin layers of sand, separated from
one another by layers still finer of coaly, or micaceous matter, that are almost exactly parallel, and

continue so to a great extent without any sensible
deviation. These planes can have acquired their
parallelism only in consequence of the property of
water just stated, by which it renders the surfaces
of the layers, which it deposits, parallel to its own
surface, and therefore parallel to one another.
Though such strata, therefore, may not now be
horizontal, they must have been so originally;
otherwise it is impossible to discover any cause for
their parallelism, or any rule by which it can have
been produced.

39. This argument for the original horizontality of the strata, is applicable to those that are now
farthest removed from that position.   Among such,
for instance, as are highly inclined, or even quite
vertical, and among those that are bent and incurvated in the most fantastical manner, as happens
more especially in the primary schisti, we observe, through all their sinuosities and inflections,
an equality of thickness and of distance among
their component laminæ.   This equality could only be produced by those laminæ having been ori-
ginally·spread out on a flat and level surface, from
which situation, therefore, they must afterwards
have been lifted up by the action of some powerful
cause, and must have suffered this disturbance
while they were yet in a certain degree flexible
and ductile.   Though the primary direction of the
force which thus elevated them must have been

from below upwards, yet it has been so combined
with the gravity and resistance of the mass to which
it was applied, as to create a lateral and oblique
thrust, and to produce those contortions of the
strata, which, when on the great scale, are among
the most striking and instructive phenomena of
geology.

40. Great additional force is given to this argument, in many cases, by the nature of the materials of which the stratified rocks are composed.
The beds of breccia and pudding-stone, for instance, are often in planes almost vertical, and, at
the same time, contain gravel-stones, and other
fragments of rock, of such a size and weight, that
they could not remain in their present position an
instant, if the cement which unites them were to
become soft ; and therefore they certainly had not
that position at the time when this cement was actually soft. This remark has been made by mineralogists who were not led to it by any system.
The judicious and indefatigable observer of the
Alps, describing the pudding-stone of Valorsine,
near the sources of the Arve, tells us, that he was
astonished to find it in beds almost vertical, a situation in which it could not possibly have been formed. " That particles," he adds, " of extreme tenuity, suspended in a fluid, might become agglutinated, and form vertical beds, is a thing that may
be conceived ; but that pieces of stone, of several

pounds weight, should have rested on the side of a perpendicular wall, till they were enveloped in a stony cement, and united into one mass, is a supposition impossible and absurd. It should be considered, therefore, as a thing demonstrated, that this pudding-stone was formed in a horizontal position, or one nearly such, and elevated after its induration. We know not," he continues, " the force by which this elevation has been effected; but it is an important step among the prodigious number of vertical beds that are to be met with in the Alps, to have found some that must certainly have been formed in a horizontal situation." *

41. Nothing can be more sound and conclusive than this reasoning ; and had the ingenious author pursued it more systematically, it must have led him to a theory of mountains very little different from that which we are now endeavouring to explain. If some of the vertical strata are proved to have been formed horizontally, there can be no reason for not extending the same conclusion to them all, even if we had not the support of the argument from the parallelism of the layers, which has been already stated.

42. The highly inclined position, and the manifold inflections of the strata, are not the only proofs of the disturbance that they have suffered,

---

* Voyages aux Alpes, Tom. II. § 690.

and of the violence with which they have been
forced up from their original place. Those interruptions of their continuity which are observed,
both at the surface and under it, are evidences of
the same fact. It is plain, that if they remained
now in the situation in which they were at first deposited, they would never appear to be suddenly
broken off. No stratum would terminate abruptly ; but, however its nature and properties might
change, it would constitute an entire and continued
rock, at least where the effects of waste and *detritus*
had not produced a separation. This, however, is
very far from being the actual condition of stratified bodies. Those that are much inclined, or that
make considerable angles with the horizontal plane,
must terminate abruptly where they come up to
the surface. Their doing so is a necessary consequence of their position, and furnishes no argument, it may be said, for their having been disturbed, different from that which has been already deduced from their inclination. There are, however,
instances of a breach of continuity in the strata,
under the surface, that afford a proof of the violence with which they have been displaced, different from any hitherto mentioned. Of this nature
are the *slips* or *shifts*, that so often perplex the
miner in his subterraneous journey, and which
change at once all those lines and bearings that had
hitherto directed his course. When his mine

reaches a certain plane, which is sometimes perpendicular, sometimes oblique to the horizon, he finds
the beds of rock broken asunder, those on the one
side of the plane having changed their place, by
sliding in a particular direction along the face of
the others.   In this motion they have sometimes
preserved their parallelism, that is, the strata on
one side of the *slip* continue parallel to those on
the other ; in other cases, the strata on each side
become inclined to one another, though their
identity is still to be recognized by their possessing
the same thickness, and the same internal characters.   These *shifts* are often of great extent, and
must be measured by the quantity of the rock
moved, taken in conjunction with the distance to
which it has been carried.   In some instances, a
vein is formed at the plane of the shift or slip, filled with materials of the kinds which will be hereafter mentioned ; in other instances, the opposite
sides of the rock remain contiguous, or have the
interval between them filled with soft and unconsolidated earth.   All these are the undeniable effects of some great convulsion, which has shaken
the very foundations of the earth ; but which, far
from being a disorder in nature, is part of a regular
system, essential to the constitution and economy
of the globe.

The production of the appearances now described, belongs, without doubt, to different periods of

time; and, where slips intersect one another, we can often distinguish the less from the more an-cient. They are all, however, of a date posterior to that at which the waving and undulated forms of the strata were acquired, as they do not carry with them any marks of the softness of the rock, but many of its complete induration.

The same phenomenon which is thus exemplifi-ed on a great scale in the bowels of the earth, is often most beautifully exhibited in single specimens of stone, and is accompanied with this remarkable circumstance, that the *integrity* of the stone is not destroyed by the shifts, whatever wounds had been made in it being healed, and the parts firmly re-united to one another. *

43. Though such marks of violence as have been now enumerated are common in some degree to all the strata, they abound most among the primary, and point out these as the part of our globe which has been exposed to the greatest vicissitudes. At their junction with the secondary, or where they emerge, as it were, from under the latter, pheno-mena occur, which mark some of those vicissitudes with astonishing precision; phenomena of which the nature was first accurately explored, and the consequences fully deduced, by the geologist whose system I am endeavouring to explain. He ob-

---

* NOTE XI.

served, in several instances, that where the primary schistus rises in beds almost vertical, it is covered by horizontal layers of secondary sandstone, which last are penetrated by the irregular tops of the schistus, and also involve fragments of that rock, some angular, others round and smooth, as if worn by attrition. From this he concluded, that the primary strata, after being formed at the bottom of the sea, in planes nearly horizontal, were raised, so as to become almost vertical, while they were yet covered by the ocean, and before the secondary strata had begun to be deposited on them. He also argued, that, as the fragments of the primary rock, included in the secondary, are many of them rounded and worn, the deposition of the latter must have been separated from the elevation of the former by such an interval of time, as gave room for the action of waste and decay, allowing those fragments first to be detached, and afterwards wrought into a round figure. *

44. Indeed, the interposition of a breccia between the primary and secondary strata, in which the fragments, whether round or angular, are always of the primary rock, is a fact so general, and the quantity of this breccia is often so great, that it leads to a conclusion more paradoxical than any of the preceding, but from which, nevertheless, it

---

* NOTE XII.

seems very difficult to withhold assent. Round gravel, when in great abundance, agreeably to a remark already made, must necessarily be considered as a production peculiar to the beds of rivers, or the shores of continents, and as hardly ever formed at great depths under the surface of the sea. It should seem, then, that the primary schistus, after attaining its erect position, had been raised up to the surface, where this gravel was formed; and from thence had been let down again to the depths of the ocean, where the secondary strata were deposited on it. Such alternate elevations and depressions of the bottom of the sea, however extraordinary they may seem, will appear to make a part of the system of the mineral kingdom, from other phenomena hereafter to be described.

45. On the whole, therefore, by comparing the actual position of the strata, their erectness, their curvature, the interruptions of their continuity, and the transverse stratification of the secondary in respect of the primary, with the regular and level situation which the same strata must have originally possessed, we have a complete demonstration of their having been disturbed, torn asunder, and moved angularly, by a force that has, in general, been directed from below upwards. In establishing this conclusion, we have reasoned more from the facts which relate to the *angular elevation* of

the strata, than from those which relate to their
absolute elevation, or their translation to a greater
distance from the centre of the earth.   This has
been done, because the appearances, which respect
the absolute lifting up of the strata are more ambiguous than those, which respect the change of
their angular position.   The former might be accounted for, could they be separated from the latter, in two ways, viz. either by the retreat of the
sea, or the raising up of the land ; but the latter
can be explained only in one way, and force us of
necessity to acknowledge the existence of an ex-
panding power, which has acted on the strata with
incredible energy, and has been directed from the
centre toward the circumference.

46. When we are assured of the existence of
such a power as this in the mineral regions, we
should argue with singular inconsistency if we did
not ascribe to it all the other appearances of motion in those regions, which it is adequate to produce.   If nature in her subterraneous abodes is
provided with a force that could burst asunder the
massy pavement of the globe, and place the fragments upright upon their edges, could she not, by
the same effort, raise them from the greatest depths
of the sea, to the highest elevation of the land ?
The cause that is adequate to one of these effects
is adequate to them both together ; for it is a principle well known in mechanical philosophy, that

the force which produces a parallel motion, may,
according to the way in which it is applied, produce also an angular motion, without any diminution of the former effect. It would, therefore, be
extremely unphilosophical to suppose, that any
other cause has changed the relative level of the
strata, and the surface of the sea, than that which
has, in so many cases, raised the strata from a horizontal to a highly inclined, or even vertical situation : it would be to introduce the action of more
causes than the phenomena require, and to forget,
that nature, whose operations we are endeavouring
to trace, combines the possession of infinite resources with the most economical application of
them.

47. From all, therefore, that relates to the position of the strata, I think I am justified in affirming, that their disturbance and removal from the
place of their original formation, by a force directed from below upwards, is a fact in the natural history of the earth, as perfectly ascertained as any
thing which is not the subject of immediate observation. As to the power by which this great effect has been produced, we cannot expect to decide
with equal evidence, but must be contented to pass
from what is certain to what is probable. We may,
then, remark, that of the forces in nature to which
our experience does in any degree extend, none
seems so capable of the effect we would ascribe to

it, as the expansive power of heat; a power to which no limits can be set, and one, which, on grounds quite independent of the elevation of the strata, has been already concluded to act with great energy in the subterraneous regions. We have, indeed, no other alternative, but either to adopt this explanation, or to ascribe the facts in question to some secret and unknown cause, though we are ignorant of its nature, and have no evidence of its existence.

We are therefore to suppose, that the power of the same subterraneous heat, which consolidated and mineralized the strata at the bottom of the sea, has since raised them up to the height at which they are now placed, and has given them the various inclinations to the horizon which they are found actually to possess.

48. The probability of this hypothesis will appear greatly increased, when it is considered, that, besides those now enumerated, there are other indications of movement among the bodies of the mineral kingdom, where effects of heat more characteristic than simple expansion are clearly to be discovered. Thus, on examining the marks of disorder and movement which are found among the strata, it cannot fail to be observed, that notwithstanding the fracture and dislocation, of which they afford so many examples, there are few empty spaces to be met with among them, as far as our ob-

servation extends. The breaches and separations are numerous, and distinct; but they are, for the most part, completely filled up with minerals of a kind quite different from the rock on each side of them, and remarkable for containing no vestiges of stratification. We are thus led to consider the unstratified minerals, the second of the divisions into which the whole mineral kingdom, viewed geologically, ought to be distinguished. These minerals are immediately connected with the disturbance of the strata, and appear, in many instances, to have been the instruments of their elevation.

SECTION II

1. *Metallic Veins.*

49. THE unstratified minerals exist either in
veins, intersecting the stratified, or in masses surrounded by them.   Veins are of various kinds, and
may in general be defined, separations in the continuity of a rock, of a determinate width, but extending indefinitely in length and depth, and filled
with mineral substances, different from the rock itself.   The mineral veins, strictly so called, are
those filled with crystallized substances, and containing the metallic ores.

That these veins are of a formation subsequent
to the hardening and consolidation of the strata
which they traverse, is too obvious to require any
proof; and it is no less clear, from the crystallized and sparry structure of the substances contained in them, that these substances must have
concreted from a fluid state.   Now, that this fluidity was simple, like that of fusion by heat, and
not compound, like that of solution in a men-

struum, is inferred from many phenomena. It is inferred from the acknowledged insolubility of the substances that fill the veins, in any one menstruum whatsoever; from the total disappearance of the solvent, if there was any; from the complete filling up of the vein by the substances which that solvent had deposited; from the entire absence of all the appearances of horizontal or gradual deposition; and, lastly, from the existence of close cavities, lined with crystals, and admitting no egress to any thing but heat.

50. To the same effect may be mentioned those groups of crystals composed of substances the most different, that are united in the same specimen, all intersecting and mutually impressing one another. These admit of being explained, on the supposition that they were originally in fusion, and became solid by the loss of heat; a cause that acted on them all alike, and alike impelled them to crystallize: But the appearances of simultaneous crystallization seem incompatible with the nature of deposition from a solvent, where, with respect to different substances, the effects must take place slowly, and in succession.

51. The metals contained in the veins which we are now treating of, appear very commonly in the form of an ore, mineralized by sulphur. Their union with this latter substance can be produced, as we know, by heat, but hardly by the way of so-

lution in a menstruum, and certainly not at all, if
that menstruum is nothing else than water. The
metals, therefore, when mineralized by sulphur,
give no countenance to the hypothesis of aqueous
solution ; and still less do they give any when they
are found native, as it is called, that is, malleable,
pure and uncombined with any other substance.
The great masses of native iron found in Siberia
and South America are well known ; and nothing
certainly can less resemble the products of a chemical precipitation.   Gold, however, the most perfect of the metals, is found native most frequently ;
the others more rarely, in proportion nearly to the
facility of their combination with sulphur.   Of all
such specimens it may be safely affirmed, that if
they have ever been fluid, or even soft, they must
have been so by the action of heat ; for, to suppose
that a metal has been precipitated, pure and uncombined from any menstruum, is to trespass
against all analogy, and to maintain a physical impossibility.   But it is certain, that many of the native metals have once been in a state of softness,
because they bear on them impressions which they
could not have received but when they were soft.
Thus, gold is often impressed by quartz and other
stones, which still adhere to it, or are involved in
it.   Specimens of quartz, containing gold and silver shooting through them, with the most beautiful and varied ramifications, are every where to be

met with in the cabinets of the curious ; and contain, in their structure, the clearest proof, that the
metal and the quartz have been both soft, and have
crystallized together. By the compactness, also,
of the body which they form, they show, that when
they acquired solidity, it was by the concretion of
the whole mass, and not by such partial concretion
as takes place when a solvent is separated from substances which it held in solution.

52. Native copper is very abundant ; and some
specimens of it have been found crystallized. Here
the crystallization of the metal is a proof that it
has passed from a fluid to a solid state ; and its
purity is a proof that it did not make that transition by being precipitated from a menstruum.

53. Again, pieces of native manganese have been
found possessing so exactly the characters peculiar
to that metal when reduced in our furnaces, that
it is impossible to consider them as deriving their
figure and solidity from any cause but fusion. The
ingenious author who describes these specimens,
Lapeyrouse, was so forcibly struck with this resemblance, that he immediately drew the same
conclusion from it which is drawn here, attributing
the only difference, which he remarked between
the native and the artificial *regulus*, to the different energy with which the same agent works when
employed by nature and by art. *

---

* Theory of the Earth, Vol. I. p. 68. Journal de Phys.
Janvier, 1786.

54. All these appearances conspire to prove, that the materials which fill the mineral veins were melted by heat, and forcibly injected, in that state, into the clefts and fissures of the strata. These fissures we must conceive to have arisen, not merely from the shrinking of the strata while they acquired hardness and solidity, but from the violence done to them, when they were heaved up and elevated in the manner which has already been explained. *

55. When these suppositions are once admitted, the other leading facts in the history of metallic veins will be readily accounted for. Thus, for instance, it is evident to what we must ascribe the fragments of the surrounding rock that are often found immersed in the veins, and encompassed on all sides by crystallized substances. These fragments being no doubt detached by the concussion, which at once tore asunder and elevated the strata, were sustained by the melted matter that flowed at the same time upward through the vein. Large masses of rock are often found in this manner completely insulated ; one of these, which M. Deluc has described with great accuracy, is no less than a vast segment of a mountain. †

56. The immense violence which has accompanied the formation of mineral veins, is particularly

---

* NOTE XIII.

† Lettres Physiques, &c. Tom. III. p. 361.

marked by the slips and shifts of the strata on each
side of them, all tending to show what mighty
changes have taken place in those regions, which
our imagination erroneously paints as the abode of
everlasting silence and rest. This shifting of the
strata is best observed, where the veins make a
transverse section of beds of rock, considerably inclined to the horizon. There it is common to see
the beds on one side of the vein slipped along from
the corresponding beds on the other side, and removed sometimes in a horizontal, sometimes in an
oblique direction. In this way, not only the strata
are shifted, but veins, which intersect one another,
are also shifted themselves. They are *heaved*, as
it is called in the significant language of the miners, and forced out of their direction. It is impossible, in such a case, but to connect in the mind
the formation of the vein, and the production of
the slips which accompany it, and to regard them
as parts of the same phenomenon.

57. Where these slips are horizontal, and exhibit great bodies of strata carried from their place,
while the parts of the transferred mass remain undisturbed relatively to one another, they furnish a
clear proof, that this change of place has not arisen
from the falling in of the roofs of caverns, as some
geologists suppose. The horizontal direction, and
the regularity of the movement, are incompatible
with the action of such a cause as this ; and in-

deed it is highly interesting to remark, in the midst
of the signs of disturbance which prevail in the
bowels of the earth, that there reigns a certain
symmetry and order, which indicate the action of a
force of incredible magnitude, but slow and gradual
in its effects.   The parts of the mass moved are undisturbed relatively to one another ; what has been
broken has been cemented ; the breaches of continuity have been filled up and healed ; and every
where we see the operation of a cause that could
unite as well as separate.   The twofold action of
heat to expand and to melt, could scarce be pointed out more clearly by any system of appearances.

58. As a long period was no doubt required for
the elevation of the strata, the rents made in them
are not all of the same date, nor the veins all of the
same formation.   This is clear in the case of one
vein producing a shift or slip in another ; for the
vein which forces the other out of its place, and
preserves its own direction, is evidently the more
recent of the two, and must have had its materials
in a state of activity, when those of the other were
inert.   Sometimes, also, at the intersection of two
veins, we may trace the current of the materials of
the one, across those of the other ; and here, of
consequence, the relative antiquity is determined
just as in the former instance.

59. The want of any appearance of stratification
in mineral veins has already been taken notice of.

11

There is, however, to be observed, in many instances, a tendency to a regular arrangement of
the substances contained in them; those of the
same kind forming coats parallel to the sides of the
vein, and nearly of an equal thickness. This phenomenon is considered as one of the strongest arguments in favour of the Neptunian system, but
has nothing in it, in the least incompatible with
that theory which ascribes the formation of veins to
the action of subterraneous heat. When melted
matter from the mineral regions was thrown up into the veins, that which was nearest to the sides
would soonest lose its heat. The similar substances, also, would unite while this process was
going forward, and would crystallize, as in other
cases of congelation, from the sides toward the interior. There is the more reason for supposing
this to have been the case, that the same sort of
coating is often observed on the inside of close cavities, which are, nevertheless, so constructed, as to
afford a demonstration that no chemical solvent
was ever included in them, (§ 74.) Some veins, it
must also be considered, may have been filled by
successive injections of melted matter, and this
would naturally give rise to a variety of separate incrustations. *

60. In the view now given of metallic veins,

---

* See some farther remarks on this subject at NOTE XIII.

they have been considered as traversing only the
stratified parts of the globe. They do, however,
occasionally intersect the unstratified parts, particularly the granite, the same vein often continuing its
course across rocks of both kinds, without suffering
any material change ; and, if we have hitherto paid
no attention to this circumstance, it is because the
order pursued in this essay required, that the relation of the veins to stratified bodies should be first
treated of. Besides, the facts in the natural history
of veins, whether contained in stratified or unstratified rocks, are so nearly alike, that in a general
view of geology, they do not require to be distinguished. It is material to remark, that, though
metallic veins are found indiscriminately in all the
different kinds of rock, whether stratified or otherwise, they are most abundant in the class of primary schisti. All the countries most remarkable
for their mines, and the mountains distinguished by
the name of metalliferous, are primary, and the instance of Derbyshire is perhaps the most considerable exception to this rule that is known. This
preference, which the metals appear to give to the
primary strata, is very consistent with Dr Hutton's
theory, which represents the rocks of that order as
being most changed from their original position,
and those on which the disturbing forces of the
subterraneous regions have acted most frequently,
and with greatest energy. The primary strata are

the lowest, also, and have the most direct communication with those regions from which the mineral
veins derive all their riches.

2. *Of Whinstone.*

61. Beside the veins filled with spar, and containing the metallic ores, the strata are intersected
by veins of whinstone, porphyry, and granite, the
characters of which are next to be examined.

The term *whin*, or *whinstone*, with Dr Hutton, like the word *trap*, with the German mineralogists, denotes a class of stones, comprehending
several distinct species, or at least varieties. The
common *basali*, the *wacken*, *mullen*, and *crag* of
Kirwan, the *grunstein* of Werner, and the *amygdaloid*, are comprehended under the name of whin.
All these stones have a tendency to a spathose
structure, and discover at least the rudiments of
crystallization. They are, at the same time, without any mark of stratification in their internal texture, as they are also, for the most part, in their
outward configuration ; and, as the different species
here enumerated compose, not unfrequently, parts
of the same continuous rock, the change from one
to another being made through a series of insensible gradations, they may safely be regarded by the
geologist as belonging to the same *genus.*

62. Whin, though not stratified, exists in two different ways, that is, either in veins, (called in Scotland *dykes*,) traversing the strata like the veins already described, or in irregular masses, incumbent on the strata, and sometimes interposed between them. In both these forms, whinstone has nearly the same characters, and bears, in all its varieties, a most striking resemblance to the lavas which have actually flowed from volcanoes on the surface of the earth. This resemblance is so great, that the two substances have been often mistaken for one another ; and many rocks, which have been pronounced to be the remains of extinguished volcanoes, by mineralogists of no inconsiderable name, have been found, on closer examination, to be nothing else than masses or veins of whinstone. This latter stone is indeed only to be distinguished from the former, by a careful examination of the internal characters of both ; and chiefly from this circumstance, that whinstone often contains calcareous spar and zeolite, whereas neither of these substances is found in such lavas, as are certainly known to have been thrown out by volcanic explosions.

Now, from these circumstances of affinity between lava and whinstone, on the one hand, and of diversity on the other, as the formation of the one is known, it should seem that some probable conclusion may be drawn concerning the formation of the other. The affinity in question is constant and es-

sential ; the difference variable and accidental ; and
this naturally leads to suspect, that the two stones
have the same origin; and that, as lava is certainly
a production of fire, so probably is whinstone.

63. But, in order to see whether this hypothesis
will explain the diversity of the two substances,
without which it will not be entitled to much attention, we must remark, that the presence of carbonate of lime in a body that has been fused, argues,
agreeably to the principles formerly explained, that
the fusion was brought about under a great compressing force, that is to say, deep in the bowels of
the earth, or in the great laboratory of the mineral
regions.   We are, therefore, to suppose that the
fusion of the whin was performed in those regions, where the compression was sufficient to preserve the carbonic gas in union with the calcareous
earth, so that these two substances melted together,
and, on cooling, crystallized into spar.   In the
lavas, again, thrown out by volcanic eruption, the
fusion, as we know, wherever it may begin, continues in the open air, where the pressure is only
that of the atmosphere : the calcareous earth, which,
therefore, may have been, in the form of a carbonate,
among the materials of this lava, must be converted
into quicklime, and become infusible ; hence the
want of calcareous spar in lavas that have flowed at
the surface.

Thus, whinstone is to be accounted a subterraneous, or *un-erupted* lava ; and our theory has the

advantage of explaining both the affinity and the difference between these stony bodies, without the introduction of any new hypothesis. In the Neptunian system, the affinity of whinstone and lava is a paradox which admits of no solution.

64. The columnar structure sometimes found in that species of whinstone called basaltes, is a fact which has given rise to much discussion ; and it must be confessed, that though one of the most striking and peculiar characters of this fossil, it is not that which gives the clearest and most direct information concerning its origin. One circumstance, however, very much in favour of the opinion that basaltic rocks owe their formation to fire, is, that the columnar form is sometimes assumed by the lava actually erupted from volcanoes. Now, it is certainly of no small importance, to have the synthetic argument on our side, and to know, that basaltic columns can be produced by fire ; though, no doubt, to give absolute certainty to our conclusion, it would be necessary to show, that there are in nature no other means but this by which these columns can be formed. This sort of evidence is hardly to be looked for ; but since the power of fusion, to produce the phenomena in question, is perfectly established, and since the production of the same phenomena in the humid way is a mere hypothesis, if there be the least reason to suspect the action of subterraneous heat as one of the causes of mineralization, every maxim of sound philosophy requires

that the basaltic structure, in all cases, should be
ascribed to it.

65. The Neptunists will no doubt allege, with
Bergman, that, in the drying of starch, clay, and a
few other substances, something analogous to basaltic columns is produced. Here, however, a most
important difference is to be remarked, corresponding very exactly to one of the characters which we
have all along observed to distinguish the products
of aqueous, from those of igneous consolidation.
The columns formed by the substances just mentioned, are distant from one another : they are separated by fissures which widen from the bottom to
the top, and which arise from the shrinking and
drying of the mass. In the basaltic columns, no
such openings, nor vacuity of any kind is found ;
the pillars are in contact, and, though perfectly distinct, are so close, that the sharp edge of a wedge
can hardly be introduced between them. This is a
great peculiarity in the basaltic structure, and is
strongly expressive of this fact, that the mass was
all fluid together, and that its parts took their new
arrangement, not in consequence of the separation
of a fluid from a solid part, by which great shrinking and much empty space might be produced ;
but in consequence of a cause which, like refrigeration, acted equally on all the parts of the mass, and
preserved their absolute contact after their fluidity
had ceased.

66. A mark of fusion, or at least of the operation of heat, which whinstone possesses in common
with many other minerals, is its being penetrated
by pyrites, a substance, as has been already remarked, that is of all others most exclusively the production of fire. Another mark of fusion, more
distinctive of whin, is, that both in veins and in
masses it sometimes includes pieces of sandstone,
or of the other contiguous strata, completely insulated, and having the appearance of fragments of
rock, floating in a fluid sufficiently dense and ponderous to sustain their weight. Though these
fragments have been too refractory to be reduced
into fusion themselves, they have not remained entirely unchanged, but are, in general, extremely
indurated, in comparison of the rock from which
they appear to have been detached.

67. Similar instances of extraordinary induration are observed in the parts of the strata in contact with whinstone, whether they form the sides
of the veins, or the floors, and roofs of the masses
into which the whinstone is distributed. The strata
whether sandy or argillaceous, in such situations,
are usually extremely hard and consolidated ; the
former in particular lose their granulated texture,
and are sometimes converted into perfect jasper.
This interesting remark was first made by Dr Hutton, and the truth of it has been verified by a great
number of subsequent observations.

68. To the same excellent geologist we are indebted for the knowledge of an analogous fact, attendant on the passage of whinstone veins through coal strata. As the beds of stone where they are in contact with veins of whin, seem to acquire additional induration, so those of coal, in like circumstances, are frequently found to have lost their fusibility, and to be reduced nearly to the condition of coke, or of charcoal. The existence of coal of this kind has been already mentioned, and considered as a proof of the operation of subterraneous heat. In the instances here referred to, that is, where the charring of the coal is limited to those parts of the strata which are in contact with the whin, or in its immediate vicinity, the heat is pointed out as residing in the vein ; and this is to be accounted for only on the supposition of the melted whin, at a period subsequent to the consolidation of the coal, having flowed through the openings of the strata. The heat has been powerful enough, in many cases, to drive off the bituminous matter of the coal, and to force it into colder and more distant parts. Few facts, in the history of fossils, are more remarkable than this, and none more directly assimilates the operations of the mineral regions, with those that take place at the surface of the earth.

69. Again, the disturbance of the strata, wherever veins of whinstone abound, if not a direct

proof of the original fluidity of the whinstone, is a
clear indication of the violence with which it was
introduced into its place.   This disturbance of the
position of the strata, by shifting, unusual elevation, and other irregularities, where they are intersected by whinstone veins, is a fact so well known
to miners, that when they meet with any sudden
change in the lying of the *metals*, they are wont to
foretell their approach to masses, or veins of unstratified matter ; and, in their figurative language,
point them out as the causes of the confusion with
which they are so generally accompanied. *   The
mineral veins likewise, as well as the strata, are often heaved and shifted by the veins of whinstone.

70. Whinstone of every species is found frequently interposed in tabular masses, between beds
of stratified rocks ; and it then adds to the indications of its igneous origin, already enumerated, some
others that are peculiar to it when in this situation.
In such instances, it is not uncommon to find the
strata in some places, contiguous to the whin, elevated, and bent with their concavity upward, so
that they appear clearly to have been acted on by a
force that proceeded from below, at the same time
that they were softened, and rendered in some degree flexible : it is needless to remark, that these

* A *Trouble* is the name which the colliers in this country give to a vein of whinstone.

effects can be explained by nothing but the fusion
of the whin ; and that the great force with which
it was impelled against the strata, could be produced by no cause but heat, acting in the manner
that is here supposed.

71. Again, if it be true that the masses of whin,
thus interposed among the strata, were introduced
there, after the formation of the latter, we might
expect to find, at least in many instances, that the
beds on which the whinstone rests, and those by
which it is covered, are exactly alike. If these
beds were once contiguous, and have been only
heaved up and separated by the irruption of a fluid
mass of subterraneous lava, their identity should
still be recognised.   Now, this is precisely what is
observed ; it is known to hold in a vast number of
instances, and is strikingly exemplified in the rock
of *Salisbury Crag*, near Edinburgh.

This similarity of the strata that cover the masses
of whinstone, to those that serve as the base on
which they rest, and again the dissimilitude of both
to the interposed mass, are facts which I think can
hardly receive any explanation, on the principles of
the Neptunian theory. If these rocks, both stratified and unstratified, are to be regarded as productions of the sea, the circumstances would require to
be pointed out, which have determined the whinstone, and the beds that are all round it, to be so
extremely unlike in their structure, though formed

at the same time, and in the immediate vicinity of
one another ; as also those circumstances, on the
other hand, which determined the stratified deposits above and below the whinstone, to be precisely the same, though the times of their formation
must have been very different.   The homogeneous
substances, thus, placed at a distance, and the heterogeneous brought so closely together, are phenomena equally unaccountable, in a theory that ascribes their origin to the operation of the same element, and that necessarily dates their formation
according to the order in which they lie, one above
another.

72. If, indeed, in these instances, the gradation
were insensible, as some have asserted it to be, between the strata and the interposed mass, so that
it was impossible to point out the line where the
one ended and the other began, whatever difficulties we might perceive in the Neptunian theory,
we should find it hard to substitute a better in its
room.   But the truth seems to be, that, in the
cases we are now treating of, no such gradation exists ; and that, though where the two kinds of rock
come into contact a change is often observed, by
the strata having acquired an additional degree of
induration, yet the line of separation is well defined, and can be precisely ascertained.   This at least
is certain, that innumerable specimens, exhibiting
such lines of separation, are to be met with ; and

wherever care has been taken to obtain a fresh
fracture of the stone, and to remove the effects of
accidental causes, even where the two rocks are
most firmly united, and most closely assimilated, I
am persuaded that no uncertainty has ever remained as to the line of their separation. For these
reasons, it seems probable that the gradual transition of basaltes into the adjoining strata, is in all
cases imaginary, and is, in truth, a mere illusion,
proceeding from hasty and inaccurate observation.

73. Another remarkable fact in the natural history of the whinstone rocks, remains yet to be
mentioned, and with it I shall conclude the argument, as far as these rocks are concerned.

Some of the species of whinstone are the common matrices of agates and chalcedonies, which lie
inclosed in them in the form of round nodules.
The original fluidity of these nodules is evinced by
their figured, and sometimes crystallized structure,
and indeed is so generally admitted, that the only
question concerning them is, whether this fluidity
was the effect of heat or of solution. To answer this
question, Dr Hutton observes, that the formation
of the concentric coats, of which the agate is usually composed, has evidently proceeded from the circumference toward the centre, the exterior coats always impressing the interior, but never the reverse.
The same thing also follows from this other fact,
that when there is any vacuity within the agate, it

is usually at the centre, and there too are found the
regular crystals, when any such have been formed.
It therefore appears certain, that the progress of
consolidation has been from the circumference in-
wards, and that the outward coats of the agate were
the first to acquire solidity and hardness.

74. Now, it must be considered that these coats
are highly consolidated ; that they are of very pure
siliceous matter, and are utterly impervious to every
substance which we know of, except light and heat.
It is plain, therefore, that whatever at any time,
during the progress of consolidation, was contained
within the coats already formed, must have remained there as long as the agate was entire, without
the least possibility of escape. But nothing is found
within the coats of the agate save its own substance ;
therefore, no extraneous substance, that is to say
no solvent, was ever included within them.   The
fluidity of the agate was therefore simple, and un-
assisted by any menstruum.

In this argument, nothing appears to me wanting, that is necessary to the perfection of a physical,
I had almost said of a mathematical, demonstration.
It seems, indeed, to be impossible that the igneous
origin of fossils could be recorded in plainer language, than by the phenomenon which has just
been described.

75. The examination of particular specimens of
agates and chalcedonies, affords many more argu-

ments of the same kind, which Dr Hutton used
to deduce with an acuteness and vivacity, which
his friends have often listened to with great admiration and delight. * These, however, must be passed over at present; and I have only further to remark, that a series of the most interesting experiments, instituted by Sir James Hall, and published
in the Transactions of the Royal Society of Edinburgh, † has removed the only remaining objection
that could be urged against the igneous origin of
whinstone. This objection is founded on the common observation, that when a piece of whinstone or
basaltes is actually melted in a crucible, on cooling,
it becomes glass, and loses its original character entirely; and from thence it was concluded, that this
character had not been originally produced by fusion. The experiments above mentioned, however,
have shown, in the most satisfactory manner, that
melted whin, by *regulated* or by slow cooling, is
prevented from assuming the appearance of glass,
and becomes a stony substance, hardly to be distinguished from whinstone or lava.

The experiments of another ingenious chemist,
Dr Kennedy, have shown, that whinstone contains
mineral alkali, by which, of course, its fusion must
have been assisted. ‡ Dr Hutton used to ascribe

---

* NOTE XIV. † Vol. V. p. 43.
‡ Trans. R. S. Edin. Vol. V. p. 85.

its fusibility, in a great measure at least, to the quantity of iron contained in it : both these causes have no doubt united to render it more easily melt-ed than the ordinary materials of the strata.

76. In a word, therefore, to conceive aright the origin of that class of unstratified rocks, distinguish-ed by the name of whinstone, we must suppose, that long after the consolidation of the strata, and during the time of their elevation, the materials of the former were melted by the force of subterrane-ous heat, and injected among the rents and fissures of the rocks already formed.   In this manner were produced the veins or dikes of whinstone ; and, where circumstances allowed the stream of melted matter to diffuse itself more widely, tabular masses were formed, which were afterwards raised up, to-gether with the surrounding strata, above the level of the sea, and have been since laid open by the operation of those causes that continually change and waste the surface of the land.

These unstratified rocks are not, however, all the work of the same period ; they differ evidently in the date of their formation, and it is not unusual, to find tabular masses of one species of whin, inter-sected by veins of another species.   Indeed, of all the fossil bodies which compose the present land, the veins of whin appear to be the most recently consolidated. *

* Note xiv.

Porphyry may so properly be regarded as a variety of whin, distinguished only by involving crystallized feldspar, that, in a geological sketch like the
present, it is hardly entitled to a separate article.
Like the other kinds of whin, it exists both in veins
and in tabular masses, having, no doubt, an origin
similar to that which has just been described. Porphyry, however, has the peculiarity of being rarely
found in any but the primary strata ; it seems to be
the whinstone of the old world, or at least that
which is of the highest antiquity in the present. It
nowhere, I believe, assumes a columnar, or basaltic
appearance, of any regularity ; but this is also true
of many other varieties of whin, of all, indeed, except the most compact and homogeneous. These
differences are not so considerable as to require our
entering into any particular detail concerning the
natural history of this fossil.

3. Granite.

77. The term Granite is used by Dr Hutton to
signify an aggregate stone, in which quartz, feldspar, and mica are found distinct from one another,
and not disposed in layers. The addition of hornblende, schorl, or garnet, to the three ingredients
just mentioned, is not understood to alter the *genus*
of the stone, but only to constitute a specific differ-

ence, which it is the business of lithology to mark by
some appropriate character, annexed to the generic
name of granite.

The fossil now defined exists, like whinstone and
porphyry, both in masses and in veins, though most
frequently in the former.   It is like them unstratified in its texture, and is regarded here, as being
also unstratified in its outward structure. *   One
ingredient which is essential to granite, namely,
quartz, is not contained in whinstone ; and this
circumstance serves to distinguish these *genera*
from one another, though, in other respects, they
seem to be united by a chain of insensible gradations, from the most homogeneous basaltes, to granite the most highly crystallized.

---

* Those rocks that consist of the ingredients here enumerated, if they have at the same time a schistose texture,
or a disposition into layers, are properly distinguished from
granite, and called Gneiss, or Granitic Schistus.   But it has
been questioned whether a stone does not exist composed of
these ingredients, and destitute of a schistose texture, but
yet divided into large beds, visible in its external form.   Dr
Hutton supposes such a stone not to exist, or at least not to
constitute any such proportion of the mineral kingdom, as to
entitle it to particular consideration, in the general speculations of geology.

Whether this supposition is perfectly correct, may require
to be farther considered : this, however, is certain, that a
rock, in all respects conformable to it, composes a great proportion of what are usually called the granite mountains
See NOTE xv.

78. Granite, it has been just said, exists most commonly in masses ; and these masses are rarely, if ever, incumbent on any other rock : they are the basis on which others rest, and seem, for the most part, to rise up from under the ancient, or primary strata. The granite, therefore, wherever it is found, is inferior to every other rock ; and as it also composes many of the greatest mountains, it has the peculiarity of being elevated the highest into the atmosphere, and sunk the deepest under the surface, of all the mineral substances with which we are acquainted.

Notwithstanding the circumstance of not being alternated with stratified bodies, which constitutes a remarkable difference between granite and whinstone, the affinity of these fossils is such as to make the similarity of their origin by no means improbable. Accordingly, in Dr Hutton's theory, granite is regarded as a stone of more recent formation than the strata incumbent on it ; as a substance which has been melted by heat, and which, when forced up from the mineral regions, has elevated the strata at the same time.

79. That granite has undergone a change from a fluid to a solid state, is evinced from the crystallized structure in which some of its component parts are usually found. This crystallization is particularly to be remarked of the feldspar, and also of the

VOL. I.   G

schorl, where there is any admixture of that substance, whether in slender spiculæ, or in larger
masses. The quartz itself is in some cases crystallized, and is so, perhaps, more frequently than is
generally supposed. The fluidity of granite, in
some former period of its existence, is so evident
from this, as to make it appear singular that it
should ever have been considered as a fossil that
had remained always the same, and one, into the
origin of which it was needless to inquire. If the
regular forms of crystallization are not to be received as proofs of the substance to which they belong
having passed from a fluid to a solid state, neither
are the figures of shells and of other supposed petrifactions, to be taken as indications of a passage from
the animal to the mineral kingdom ; so that there
is an end of all geological theories, and of all reasonings concerning the ancient condition of the
globe. To an argument which strikes equally at
the root of all theories, it belongs not to this, in particular, to make any reply.

80. We shall, therefore, consider it as admitted,
that the materials of the granite were originally
fluid ; and, in addition to this, we think it can easily be proved, that this fluidity was not that of the
elements taken separately, but of the entire mass.
This last conclusion follows, from the structure of
those specimens, where one of the substances is im-

pressed by the forms which are peculiar to another. Thus, in the Portsoy granite, * which Dr Hutton has so minutely described, the quartz is impressed by the rhomboidal crystals of the feldspar, and the stone thus formed is compact and highly consolidated. Hence, this granite is not a congeries of parts, which, after being separately formed, were somehow brought together and agglutinated; but it is certain that the quartz, at least, was fluid when it was moulded on the feldspar. In other granites, the impressions of the substances on one another are observed in a different order, and the quartz gives its form to the feldspar. This, however, is more unusual; the quartz is commonly the substance which has received the impressions of all the rest; and the spiculæ of schorl often shoot both across it and the feldspar.

The ingredients of granite were therefore fluid when mixed, or at least when in contact with one another. Now, this fluidity was not the effect of solution in a menstruum; for, in that case, one kind of crystal ought not to impress another, but each of them should have its own peculiar shape.

81. The perfect consolidation of many granites, furnishes an argument to the same effect. For, agreeably to what was already observed, in treating of the strata, a substance, when crystallizing, or

* Theory of the Earth, Vol. I. p. 104.

passing from a fluid to a solid state, cannot be free
from porosity, much less fill up completely a space
of a given form, if, at the same time, any solvent is
separated from it ; because the solvent so separated
would still occupy a certain space, and, when removed by evaporation or otherwise, would leave
that space empty.   The perfect adjustment, therefore, of the shape of one set of crystallizing bodies,
to the shape of another set, as in the Portsoy granite, and their consolidation into one mass, is as
strong a proof as could be desired, that they crystallized from a state of simple fluidity, such as, of
all known causes, heat alone is able to produce.

82. This conclusion, however, does not rest on
a single class of facts.   It has been observed in
many instances, that where granite and stratified
rocks, such as primary schistus, are in contact, the
latter are penetrated by veins of the former, which
traverse them in various directions.   These veins
are of different dimensions, some being of the
breadth of several yards, others of a few inches, or
even tenths of an inch ; they diminish as they recede from the main body of the granite, to which
they are always firmly united, constituting, indeed,
a part of the same continued rock.

These phenomena, which were first distinctly observed by Dr Hutton, are of great importance in
geology, and afford a clear solution of the two chief
questions concerning the relation between granite

and schistus. As every vein must be of a date posterior to the body in which it is contained, it follows, that the schistus was not super-imposed on the granite, after the formation of this last. If it be argued, that these veins, though posterior to the schisti, are also posterior to the granite, and were formed by the infiltration of water in which the granite was dissolved or suspended ; it may be replied, 1*mo*, That the power of water to dissolve granite, is a postulatum of the same kind that we have so often, and for such good reason, refused to concede ; and, 2*do*, That in many instances the veins proceed from the main body of the granite *upwards* into the schistus ; so that they are in planes much elevated in respect of the horizon, and have a direction quite opposite to that which the hypothesis of infiltration requires. It remains certain, therefore, that the whole mass of granite, and the veins proceeding from it, are coëval, and both of later formation than the strata.

Now, this being established, and the fluidity of the veins, when they penetrated into the schistus, being obvious, it necessarily follows, that the whole granite mass was also fluid at the same time. But this can have been brought about only by subterraneous heat, which also impelled the melted matter against the superincumbent strata, with such force as to raise them from their place, and to give them that highly inclined position in which they are still supported by the granite, after its fluidity has ceased. Thus a conclusion, rendered probable by the crystallization of granite, is established beyond all contradiction by the phenomena of granitic veins. *

83. With the granite, we shall consider the proof of the igneous origin of all mineral substances as completed. These substances, therefore, whether stratified or unstratified, owe their consolidation to the same cause, though acting with different degrees of energy. The stratified have been in general only softened or penetrated by melted matter, whereas the unstratified have been reduced into perfect fusion.

84. In this general conclusion we may distinguish two parts, which, in their degree of certainty, differ perhaps somewhat from one another. The first of these, and that which stands highest in point of evidence, consists of two propositions; namely, that the fluidity which preceded the consolidation of mineral substances was SIMPLE, that is, it did not arise from the combination of these substances with any solvent; and, next, that after consolidation, these bodies have been raised up by an expansive force acting from below, and have by that means been brought into their present situation. These two propositions seem to me to be

---

* NOTE XV.

4

supported by all the evidence that is necessary to constitute the most perfect demonstration.

85. The other part of the general conclusion, that fire, or more properly heat, was the cause of the fluidity of these mineral bodies, and also of their subsequent elevation, is not perhaps to be considered as a truth so fully demonstrated as the two preceding propositions ; it is, no doubt, a matter of THEORY ; or a portion of one of those invisible chains by which men seek to connect in the mind the state of nature that is present, with the states of it that are past ; and participates of that uncertainty from which our reasonings concerning such causes as are not direct objects of perception, are hardly ever exempted. That it participates of this uncertainty in a very slight degree, will, however, be admitted, when it is considered that the cause assigned has been proved sufficient for the effect ; that the same is not true of any other known cause ; and that this theory accounts, with singular simplicity and precision, for a system of facts so various and complex, as that which is presented by the natural history of the globe.

86. Neither can it be said that the existence of subterraneous heat is a principle assumed without any evidence, but that of the geological facts which it is intended to explain : on the contrary, it is proved by phenomena within the circle of ordinary experience, namely, those of hot springs, volcanoes,

and earthquakes. These leave no doubt of the
existence of heat, and of a moving and expansive
power, in the bowels of the earth ; so that the only questions are, at what depth is this power lodged ? to what extent, and with what intensity, does
it act ? That it is lodged at a very considerable
depth, is rendered probable by the permanency of
some of the preceding phenomena : from the earliest times many fountains have retained their heat
to the present day; and volcanoes, though they
become extinguished at length, have a very long
period allotted for their duration. The cause of
earthquakes is certainly a force that resides very
deep under the surface, otherwise the extent of the
concussion could not be such as has been observed
in many instances.

87. The intensity of volcanic fire, is another
circumstance that favours the opinion of its being
seated deep under the surface. That this intensity is considerable, is certain from the experiments
made by Sir James Hall on the fusibility of whinstone and lava ; from which it appears, that the
lowest temperature in which either of these stones
melt, is about 30^0 of Wedgewood's pyrometer.
Some mineralogists have indeed affirmed, that lava
is melted, not by the intensity of the heat applied
to it, but in consequence of a certain combination
formed between it and bituminous substances, in a
manner which they do not attempt to explain, and

which has indeed no analogy to any thing that is known. That a hypothesis, formed in such direct opposition to the most obvious principles of inductive reasoning, should have been imagined by a philosopher who had examined the phenomena of Ætna and Vesuvius with much attention, and described them with great accuracy and truth, is more wonderful than that it should have been adopted by mineralogists, whose views of nature may have been confined within a cabinet or a laboratory. It is, however, a hypothesis, which, having never had any support but from other hypotheses, hardly merited the direct refutation that it has received from the experiments just mentioned.

88. But, if the intensity of volcanic heat be such as is here stated, it will be found very difficult to account for a fire of such activity, and of such long continuance in the same spot, by any decomposition of mineral substances near the surface. In the place where this combustion is supposed to exist, it must be remembered, that there is no fresh supply of materials to replace those that have been consumed, and that, therefore, the original accumulation of these materials in one spot, must have been very unlike any thing that has ever been observed concerning the disposition of minerals in the bowels of the earth.

89. If, on the other hand, we ascribe the phenomena of volcanoes to the central heat, the account

that may be given of them is simple, and consistent
with itself.   According to all the appearances from
which the existence of such heat has been inferred
above, it is of a nature so far different from ordinary
fire, that it may require no circulation of air, and no
supply of combustible materials to support it.   It
is not accompanied with inflammation or combustion, the great pressure preventing any separation
of parts in the substances on which it acts, and the
absence of that elastic fluid without which heat
seems to have no power to decompose bodies, even
the most combustible, contributing to the unalterable nature of all the substances in the mineral regions.   There, of consequence, the only effects of
heat are fusion and expansion ; and that which
forms the nucleus of the globe may therefore be a
fluid mass, melted, but unchanged by the action of
heat.

90. If, from the confines of this nucleus, we conceive certain fissures and openings to traverse the
solid crust, and to issue at the surface of the earth,
the vapours ascending through these may in time
heat the sides of the tubes through which they pass
to a vast distance from the lower extremities.   It is,
indeed, difficult to fix the limit to which this distance may extend, on account of the great difference
between the rate at which heat moves when it has
a fluid for its vehicle, and when it is left to make
its way alone through a solid body.   In the present

case, the supply of heat is rapid, as being made by a vapour ascending through a tube of solid rock ; and the dissipation of it slow, as arising from its transmission through the rock. The waste of heat is therefore small, compared with the supply, and grows smaller at every given point, the longer the stream of heated vapour has continued to flow. Such a stream, therefore, though it may at first be condensed within a small distance of its source, will in time reach higher and higher, and may at last be able to carry its heat to an immense distance from the place of its original derivation. Thus, it is easy to conceive, that vapours from the mineral regions may convey their heat to reservoirs of water near the surface of the earth, and may in that manner produce hot springs, and even boiling fountains, like those of Reikum and Geyser.

91. When, instead of a heated vapour, melted matter is thrown up through the *shafts* or *tubes*, which thus communicate with the mineral regions, veins of whinstone and basaltes are formed in the interior of the earth. When the melted matter reaches to the surface, it is thrown out in the form of lava, and all the other phenomena of volcanoes are produced.

Lastly, Where melted matter of this kind, or vapours without being condensed, have their progress obstructed, those dreadful concussions are produced, which seem to threaten the existence even of the earth itself. Though terrible, therefore, to the present inhabitants of the globe, the earthquake has its place in the great system of geological operations, and is part of a series of events, essential, as will more clearly appear hereafter, to the general order, and to the preservation of the whole.

Such, according to this theory, are the changes which have befallen mineral substances in the bowels of the earth ; and though different for the stratified and unstratified parts of those substances, they are connected together by the same *principle*, or explained by the same *cause*. It remains to consider that part of the history of both which describes their changes after their elevation to the surface ; and here we shall find new causes introduced, which are more directly the subjects of observation, than those hitherto treated of ; causes, also, which act on all fossils alike, and alike prepare them for their ultimate destination.

SECTION III.

OF THE PHENOMENA COMMON TO STRATIFIED AND UNSTRATIFIED BODIES.

92. THE series of changes which fossil bodies are destined to undergo, does not cease with their elevation above the level of the sea ; it assumes, however, a new direction, and from the moment that they are raised up to the surface, is constantly exerted in reducing them again under the dominion of the ocean. The solidity is now destroyed which was acquired in the bowels of the earth ; and as the bottom of the sea is the great laboratory, where loose materials are mineralized and formed into stone, the atmosphere is the region where stones are decomposed, and again resolved into earth.

This decomposition of all mineral substances, exposed to the air, is continual, and is brought about by a multitude of agents, both chemical and mechanical, of which some are known to us, and many, no doubt, remain to be discovered. Among the various aëriform fluids which compose our atmosphere, one is already distinguished as the grand principle of mineral decomposition ; the others are

not inactive, and to them we must add moisture,
heat, and perhaps light; substances which, from
their affinities to the elements of mineral bodies,
have a power of entering into combination with
them, and of thus diminishing the forces by which
they are united to one another. By the action of
air and moisture, the metallic particles, particularly the iron, which enters in great abundance into
the composition of almost all fossils, becomes oxy-
dated in such a degree as to lose its tenacity; so
that the texture of the surface is destroyed, and a
part of the body resolved into earth.

93. Some earths, again, such as the calcareous,
are immediately dissolved by water; and though
the quantity so dissolved be extremely small, the
operation, by being continually renewed, produces a
slow but perpetual corrosion, by which the greatest
rocks must in time be subdued. The action of
water in destroying hard bodies into which it has
obtained entrance, is much assisted by the vicissitudes of heat and cold, especially when the latter
extends as far as the point of congelation; for the
water, when frozen, occupies a greater space than
before, and if the body is compact enough to refuse room for this expansion, its parts are torn a-
sunder by a repulsive force acting in every direction.

94. Besides these causes of mineral decomposition, the action of which we can in some measure

trace, there are others known to us only by their effects.

We see, for instance, the purest rock crystal affected by exposure to the weather, its lustre tarnished, and the polish of its surface impaired, but we know nothing of the power by which these operations are performed. Thus also, in the precautions which the mineralogist takes to preserve the fresh fracture of his specimens, we have a proof how indiscriminately all the productions of the fossil kingdom are exposed to the attacks of their unknown enemies, and we perceive how difficult it is to delay the beginnings of a process which no power whatever can finally counteract.

95. The mechanical forces employed in the disintegration of mineral substances, are more easily marked than the chemical. Here again water appears as the most active enemy of hard and solid bodies ; and, in every state, from transparent vapour to solid ice, from the smallest rill to the greatest river, it attacks whatever has emerged above the level of the sea, and labours incessantly to restore it to the deep. The parts loosened and disengaged by the chemical agents, are carried down by the rains, and, in their descent, rub and grind the superficies of other bodies. Thus water, though incapable of acting on hard substances by direct attrition, is the cause of their being so acted on ; and, when it descends in torrents, carrying with it

sand, gravel, and fragments of rock, it may be truly said to turn the forces of the mineral kingdom
against itself. Every separation which it makes is
necessarily permanent, and the parts once detached can never be united, save at the bottom of the
ocean.

96. But it would far exceed the limits of this
sketch, to pursue the causes of mineral decomposition through all their forms. It is sufficient to remark, that the consequence of so many minute, but
indefatigable agents, all working together, and having *gravity* in their favour, is a system of universal
decay and degradation, which may be traced over
the whole surface of the land, from the mountain
top to the sea shore. That we may perceive the
full evidence of this truth, one of the most important in the natural history of the globe, we will begin our survey from the latter of these stations,
and retire gradually toward the former.

97. If the coast is bold and rocky, it speaks a
language easy to be interpreted. Its broken and
abrupt contour, the deep gulfs and salient pro-
montories by which it is indented, and the proportion which these irregularities bear to the force of
the waves, combined with the inequality of hardness
in the rocks, prove, that the present line of the
shore has been determined by the action of the sea.
The naked and precipitous cliffs which overhang
the deep, the rocks hollowed, perforated, as they

are farther advanced in the sea, and at last insulated, lead to the same conclusion, and mark very
clearly so many different stages of decay. It is
true, we do not see the successive steps of this progress exemplified in the states of the same individual rock, but we see them clearly in different individuals ; and the conviction thus produced, when
the phenomena are sufficiently multiplied and varied,
is as irresistible, as if we saw the changes actually
effected in the moment of observation.

On such shores, the fragments of rock once detached, become instruments of further destruction,
and make a part of the powerful artillery with
which the ocean assails the bulwarks of the land :
they are impelled against the rocks, from which they
break off other fragments, and the whole are thus
ground against one another ; whatever be their
hardness, they are reduced to gravel, the smooth
surface and round figure of which, are the most
certain proofs of a *detritus* which nothing can resist.

98. Again, where the sea coast is flat, we have
abundant evidence of the degradation of the land
in the beaches of sand and small gravel ; the sand
banks and shoals that are continually changing ;
the alluvial land at the mouths of the rivers ; the
bars that seem to oppose their discharge into the
sea, and the shallowness of the sea itself. On such
coasts, the land usually seems to gain upon the sea,

whereas, on shores of a bolder aspect, it is the sea
that generally appears to gain upon the land.
What the land acquires in extent, however, it loses
in elevation; and, whether its surface increase or
diminish, the depredations made on it are in both
cases evinced with equal certainty.

99. If we proceed in our survey from the shores,
inland, we meet at every step with the fullest evidence of the same truths, and particularly in the
nature and economy of rivers.   Every river appears
to consist of a main trunk, fed from a variety of
branches, each running in a valley proportioned to
its size, and all of them together forming a system
of vallies, communicating with one another, and
having such a nice adjustment of their declivities,
that none of them join the principal valley, either
on too high or too low a level; a circumstance
which would be infinitely improbable, if each of
these vallies were not the work of the stream that
flows in it.

If indeed a river consisted of a single stream,
without branches, running in a straight valley, it
might be supposed that some great concussion, or
some powerful torrent, had opened at once the
channel by which its waters are conducted to the
ocean; but, when the usual form of a river is considered, the trunk divided into many branches,
which rise at a great distance from one another,
and these again subdivided into an infinity of smal-

ler ramifications, it becomes strongly impressed upon the mind, that all these channels have been cut
by the waters themselves ; that they have been
slowly dug out by the washing and erosion of the
land ; and that it is by the repeated touches of the
same instrument, that this curious assemblage of
lines has been engraved so deeply on the surface of
the globe.

100. The changes which have taken place in
the courses of rivers, are also to be traced, in
many instances, by successive platforms, of flat alluvial land, rising one above another, and marking
the different levels on which the river has run at
different periods of time. Of these, the number
to be distinguished, in some instances, is not less
than four, or even five ; and this necessarily carries
us back, like all the operations we are now treating
of, to an antiquity extremely remote : for, if it be
considered, that each change which the river makes
in its bed, obliterates at least a part of the monuments of former changes, we shall be convinced,
that only a small part of the progression can leave
any distinct memorial behind it, and that there is
no reason to think, that, in the part which we see,
the beginning is included. *

101. In the same manner, when a river under-
mines its banks, it often discovers deposits of

sand and gravel, that have been made when it ran
on a higher level than it does at present. In other
instances, the same strata are seen on both the
banks, though the bed of the river is now sunk
deep between them, and perhaps holds as winding
a course through the solid rock, as if it flowed a-
long the surface ; a proof that it must have begun
to sink its bed, when it ran through such loose materials as opposed but a very inconsiderable resistance to its stream.   A river, of which the course
is both serpentine and deeply excavated in the
rock, is among the phenomena, by which the slow
waste of the land, and also the cause of that waste,
are most directly pointed out.

102. It is, however, where rivers issue through
narrow defiles among mountains, that the identity
of the strata on both sides is most easily recognised,
and remarked at the same time with the greatest
wonder.   On observing the Potomack, where it
penetrates the ridge of the Allegany mountains, or
the Irtish, as it issues from the defiles of Altai,
there is no man, however little addicted to geological speculations, who does not immediately acknowledge, that the mountain was once continued quite
across the space in which the river now flows ; and,
if he ventures to reason concerning the cause of so
wonderful a change, he ascribes it to some great
convulsion of nature, which has torn the mountain
asunder, and opened a passage for the waters.   It

is only the philosopher, who has deeply meditated on the effects which action long continued is able to produce, and on the simplicity of the means which nature employs in all her operations, who sees in this nothing but the gradual working of a stream, that once flowed over the top of the ridge which it now so deeply intersects, and has cut its course through the rock, in the same way, and almost with the same instrument, by which the lapidary divides a block of marble or granite.

103. It is highly interesting to trace up, in this manner, the action of causes with which we are familiar, to the production of effects, which at first seem to require the introduction of unknown and extraordinary powers ; and it is no less interesting to observe, how skilfully nature has balanced the action of all the minute causes of waste, and rendered them conducive to the general good. Of this we have a most remarkable instance, in the provision made for preserving the soil, or the coat of vegetable mould, spread out over the surface of the earth. This coat, as it consists of loose materials, is easily washed away by the rains, and is continually carried down by the rivers into the sea. This effect is visible to every one ; the earth is removed not only in the form of sand and gravel, but its finer particles suspended in the waters, tinge those of some rivers continually, and those of all occasionally, that is, when they are flooded or swollen

with rains. The quantity of earth thus carried down, varies according to circumstances; it has been computed, in some instances, that the water of a river in a flood, contains earthy matter suspended in it, amounting to more than the two hundred and fiftieth part of its own bulk. * The soil, therefore, is continually diminished, its parts being transported from higher to lower levels, and finally delivered into the sea. But it is a fact, that the soil, notwithstanding, remains the same in quantity, or at least nearly the same, and must have done so, ever since the earth was the receptacle of animal or vegetable life. The soil, therefore, is augmented from other causes, just as much, at an average, as it is diminished by that now mentioned; and this augmentation evidently can proceed from nothing but the constant and slow disintegration of the rocks. In the permanence, therefore, of a coat of vegetable mould on the surface of the earth, we have a demonstrative proof of the continual destruction of the rocks; and cannot but admire the skill, with which the powers of the many chemical and mechanical agents employed in this complicated work, are so adjusted, as to make the supply and the waste of the soil exactly equal to one another.

104. Before we take leave of the rivers and the

* See Lehman, Traités de Phys. &c. Tom. III. p. 359. Note.

plains, we must remark another fact, often observed in the natural history of the latter, and clearly
evincing the former existence of immense bodies of
strata, in situations from which they have now entirely disappeared. The fact here alluded to is,
the great quantity of round and hard gravel, often
to be met with in the soil, under such circumstances, as prove, that it can only have come from
the decomposition of rocks, that once occupied the
very ground over which this gravel is now spread.
In the chalk country, for instance, about London,
the quantity of flints in the soil is every where
great ; and, in particular situations, nothing but
flinty gravel is found to a considerable depth.
Now, the source from which these flints are derived is quite evident, for they are precisely the same
with those contained in the chalk beds, wherever
these last are found undisturbed, and from the destruction of such beds they have no doubt originated.   Hence a great thickness of chalk must have
been decomposed, to yield the quantity of flints
now in the soil of these countries ; for the flints are
but thinly scattered through the native chalk, compared with their abundance in the loose earth. To
afford, for example, such a body of flinty gravel as
is found about Kensington, what an enormous
quantity of chalk rock must have been destroyed ?

105. This argument, which Dr Hutton has applied particularly to the chalk countries, may be ex-

tended to many others.   The great plain of Crau,
near the mouth of the Rhone, is well known, and
was regarded with wonder, even in ages when the
natural history of the globe was not an object of
much attention.   The immense quantity of large
round gravel-stones, with which this extensive plain
is entirely covered, has been supposed, by some mineralogists, to have been brought down by the Durance, and other torrents, from the Alps ; but, on
further examination, has been found to be of the
same kind that is contained in certain horizontal
layers of pudding-stone, which are the basis of the
whole plain.   It cannot be doubted, therefore, that
the vast body of gravel spread over it, has originated
from the destruction of layers of the same rock,
which may perhaps have risen to a great height
above what is now the surface.   Indeed, from
knowing the depth of the gravel that covers the
plain, and the average quantity of the like gravel contained in a given thickness of rock, one
might estimate how much of the latter has been
actually worn away.   Whether data precise enough
could be found, to give any weight to such a computation, must be left for future inquiry to determine. *

106. In these instances, chalk and pudding-stone,
by containing in them parts infinitely less destruc-

* NOTE XVII.

tible than their general mass, have, after they are worn away, left behind them very unequivocal marks of their existence. The same has happened in the case of mineral veins, where the substances least subject to dissolution have remained, and are scattered at a great distance from their native place. Thus gold, the least liable to decomposition of all the metals, is very generally diffused through the earth, and is found, in a greater or less abundance, in the sand of almost all rivers. But the native place of this mineral is the solid rock, or the veins and cavities contained in the rock, and from thence it must have made its way into the soil. This, therefore, is another proof of the vast extent to which the degradation of the land, and of the rock, which is the basis of it, has been carried ; and consequently, of the great difference between the elevation and shape of the earth's surface in the present, and in former ages.

107. The veins of tin furnish an argument of the same kind. The ores of this metal are very indestructible, and little subject to decomposition, so that they remain very long in the ground without change. Where there are tin veins, as in Cornwall, the tin-stone or tin ore is found in great abundance in such vallies and streams as have the same direction with the veins ; and hence the *streaming*, as it is called, or washing of the earth, to obtain the tin-stone from it. Now, if it be con-

sidered, that none of this ore can have come into
the soil but from parts of a vein actually destroyed,
it must appear evident that a great waste of these
veins has taken place, and consequently of the
schistus or granite in which they are contained.

108. These lessons, which the geologist is taught
in flat and open countries, become more striking,
by the study of those Alpine tracts, where the surface of the earth attains its greatest elevation.   If
we suppose him placed for the first time in the
midst of such a scene, as soon as he has recovered
from the impression made by the novelty and mag-
nificence of the spectacle before him, he begins to
discover the footsteps of time, and to perceive, that
the works of nature, usually deemed the most permanent, are those on which the characters of vicissitude are most deeply imprinted.   He sees himself in the midst of a vast ruin, where the precipices
which rise on all sides with such boldness and asperity, the sharp peaks of the granite mountains, and
the huge fragments that surround their bases, do
but mark so many epochs in the progress of decay,
and point out the energy of those destructive causes,
which even the magnitude and solidity of such
great bodies have been unable to resist.

109. The result of a more minute investigation,
is in perfect unison with this general impression.
Whence is it, that the elevation of mountains is so
obviously connected with the hardness and inde-

structibility of the rocks which compose them? Why
is it, that a lofty mountain of soft and secondary
rock is nowhere to be found; and that such chains,
as the Pyrenees or the Alps, never consist of any
but the hardest stone, of granite for instance, or of
those primary strata, which, if we are to credit
the preceding theory, have been twice heated in the
fires, and twice tempered in the waters, of the mineral regions? Is it not plain that this arises, not
from any direct connection between the hardness of
stones, and their height in the atmosphere, but
from this, that the waste and *detritus* to which all
things are subject, will not allow soft and weak substances to remain long in an exposed and elevated
situation? Were it not for this, the secondary
rocks, being in position superincumbent on the primary, ought to be the highest of the two, and should
cover the primary, (as they no doubt have at one
time done,) in the highest as well as the lowest situations, or among the mountains as well as in the
plains.

110. Again, wherefore is it, that among all
mountains, remarkable for their ruggedness and
asperity, the rock, on examination, is always found
of very unequal destructibility, some parts yielding
to the weather, and to the other causes of disintegration, much more slowly than the rest, and having strength sufficient to support themselves, when
left alone, in slender pyramids, bold projections,

and overhanging cliffs ? Where, on the other hand, the rock wastes uniformly, the mountains are similar to one another ; their swells and slopes are gentle, and they are bounded by a waving and continuous surface. The intermediate degrees of resistance which the rocks oppose to the causes of destruction, produce intermediate forms. It is this which gives to the mountains, of every different species of rock, a different habit and expression, and which, in particular, has imparted to those of granite that venerable and majestic character, by which they rarely fail to be distinguished.

111. The structure of the vallies among mountains, shows clearly to what cause their existence is to be ascribed. Here we have first a large valley, communicating directly with the plain, and winding between high ridges of mountains, while the river in the bottom of it descends over a surface, remarkable, in such a scene, for its uniform declivity. Into this, open a multitude of transverse or secondary vallies, intersecting the ridges on either side of the former, each bringing a contribution to the main stream, proportioned to its magnitude ; and, except where a cataract now and then intervenes, all having that nice adjustment in their levels, (99.) which is the more wonderful, the greater the irregularity of the surface. These secondary vallies have others of a smaller size opening into them ; and, among mountains of the first order,

where all is laid out on the greatest scale, these
ramifications are continued to a fourth, and even a
fifth, each diminishing in size as it increases in elevation, and as its supply of water is less. Through
them all, this law is in general observed, that where
a higher valley joins a lower one, of the two angles
which it makes with the latter, that which is obtuse is always on the descending side ; a law that
is the same with that which regulates the conflu-
ence of streams running on a surface nearly of uniform inclination. This alone is a proof that the
vallies are the work of the streams ; and indeed
what else but the water itself, working its way
through obstacles of unequal resistance, could have
opened or kept up a communication between the
inequalities of an irregular and alpine surface ?

112. Many more arguments, all leading to the
same conclusion, may be deduced from the general
facts, known in the natural history of mountains ;
and, if the Oreologist would trace back the progress of waste, till he come in sight of that original
structure, of which the remains are still so vast, he
perceives an immense mass of solid rock, naked and
unshapely, as it first emerged from the deep, and
incomparably greater than all that is now before
him. The operation of rains and torrents, modified by the hardness and tenacity of the rock, has
worked the whole into its present form ; has hollowed out the vallies, and gradually detached the

mountains from the general mass, cutting down
their sides into steep precipices at one place, and
smoothing them into gentle declivities at another.
From this has resulted a transportation of materi
als, which, both for the quantity of the whole, and
the magnitude of the individual fragments, must
seem incredible to every one, who has not learned
to calculate the effects of continued action, and to
reflect, that length of time can convert accidental
into steady causes.   Hence fragments of rock, from
the central chain, are found to have travelled into
distant vallies, even where many inferior ridges in-
tervene : hence the granite of Mont Blanc is seen
in the plains of Lombardy, or on the sides of Jura ;
and the ruins of the Carpathian mountains lie scattered over the shores of the Baltic. *

113. Thus, with Dr Hutton, we shall be disposed to consider those great chains of mountains,
which traverse the surface of the globe, as cut out
of masses vastly greater, and more lofty than any
thing that now remains.   The present appearances
afford no data for calculating the original magnitude of these masses, or the height to which they
may have been elevated.   The nearest estimate we
can form is, where a chain or group of mountains,
like those of Rosa in the Alps, is horizontally stratified, and where, of consequence, the undisturbed

---

* NOTE XVIII.

position of the mineral beds enables us to refer the
whole of the present inequalities of the surface to
the operation of waste or decay. These mountains,
as they now stand, may not inaptly be compared to
the pillars of earth which workmen leave behind
them, to afford a measure of the whole quantity of
earth which they have removed. As the pillars,
(considering the mountains as such,) are in this case
of less height than they originally were, so the measure furnished by them is but a limit, which the
quantity sought must necessarily exceed.

114. Such, according to Dr Hutton's theory,
are the changes which the daily operations of waste
have produced on the surface of the globe. These
operations, inconsiderable if taken separately, become great, by conspiring all to the same end,
never counteracting one another, but proceeding,
through a period of indefinite extent, continually
in the same direction. Thus every thing descends,
nothing returns upward ; the hard and solid bodies every where dissolve, and the loose and soft no
where consolidate. The powers which tend to
preserve, and those which tend to change the condition of the earth's surface, are never *in equilibrio ;* the latter are, in all cases, the most powerful, and, in respect of the former, are like *living* in
comparison of *dead* forces. Hence the law of decay is one which suffers no exception : The elements of all bodies were once loose and unconnect-

ed, and to the same state nature has appointed that
they should all return.

115. It affords no presumption against the reality of this progress, that, in respect of man, it is
too slow to be immediately perceived : The utmost
portion of it to which our experience can extend,
is evanescent, in comparison with the whole, and
must be regarded as the momentary increment of
a vast progression, circumscribed by no other limits
than the duration of the world. TIME performs
the office of *integrating* the infinitesimal parts of
which this progression is made up ; it collects them
into one sum, and produces from them an amount
greater than any that can be assigned.

116. While on the surface of the earth so much
is every where going to decay, no new production
of mineral substances is found in any region accessible to man. The instances of what are called
petrifactions, or the formation of stony substances
by means of water, which we sometimes observe,
whether they be ferruginous concretions, or calcareous, or, as happens in some rare cases, siliceous
stalactites, are too few in number, and too inconsiderable in extent, to be deemed material exceptions to this general rule. The bodies thus generated, also, are no sooner formed, than they become subject to waste and dissolution, like all the
other hard substances in nature ; so that they but
retard for a while the progress by which they are

all resolved into dust, and sooner or later committed to the bosom of the deep.

117. We are not, however, to imagine, that there is nowhere any means of repairing this waste ; for, on comparing the conclusion at which we are now arrived, viz. that the present continents are all going to decay, and their materials descending into the ocean, with the proposition first laid down, that these same continents are composed of materials which must have been collected from the decay of former rocks, it is impossible not to recognise two corresponding steps of the same progress ; of a progress, by which mineral substances are subjected to the same series of changes, and alternately wasted away and renovated. In the same manner, as the present mineral substances derive their origin from substances similar to themselves ; so, from the land now going to decay, the sand and gravel forming on the sea shore, or in the beds of rivers ; from the shells and corals, which in such enormous quantities are every day accumulated in the bosom of the sea ; from the drift wood, and the multitude of vegetable and animal remains continually deposited in the ocean : from all these we cannot doubt, that strata are now forming in those regions, to which nature seems to have confined the powers of mineral reproduction ; from which, after being consolidated, they are again destined to emerge,

and to exhibit a series of changes similar to the past. *

118. How often these vicissitudes of decay and renovation have been repeated, is not for us to determine : they constitute a series, of which, as the author of this theory has remarked, we neither see the beginning nor the end ; a circumstance that accords well with what is known concerning other parts of the economy of the world. In the continuation of the different species of animals and vegetables that inhabit the earth, we discern neither a beginning nor an end ; and, in the planetary motions, where geometry has carried the eye so far both into the future and the past, we discover no mark, either of the commencement or the termination of the present order. † It is unreasonable, indeed, to suppose, that such marks should any where exist. The Author of nature has not given laws to the universe, which, like the institutions of men, carry in themselves the elements of their own destruction. He has not permitted, in his works, any symptom of infancy or of old age, or any sign by which we may estimate either their future or their past duration. He may put an end, as he no doubt gave a beginning, to the present system, at some determinate period ; but we may safely

* NOTE XIX.   † NOTE XX.

conclude, that this great *catastrophe* will not be brought about by any of the laws now existing, and that it is not indicated by any thing which we perceive.

119. To assert, therefore, that, in the economy of the world, we see no mark, either of a beginning or an end, is very different from affirming, that the world had no beginning, and will have no end. The first is a conclusion justified by common sense, as well as sound philosophy ; while the second is a presumptuous and unwarrantable assertion, for which no reason from experience or analogy can ever be assigned. Dr Hutton might, therefore, justly complain of the uncandid criticism, which, by substituting the one of these assertions for the other, endeavoured to load his theory with the re-proach of atheism and impiety. Mr Kirwan, in bringing forward this harsh and ill-founded cen-sure, was neither animated by the spirit, nor guid-ed by the maxims of true philosophy. By the spi-rit of philosophy, he must have been induced to reflect, that such poisoned weapons as he was pre-paring to use, are hardly ever allowable in scienti-fic contest, as having a less direct tendency to over-throw the system, than to hurt the person of an ad-versary, and to wound, perhaps incurably, his mind, his reputation, or his peace. By the maxims of philosophy, he must have been reminded, that, in no part of the history of nature, has any mark been

discovered, either of the beginning or the end of
the present *order ;* and that the geologist sadly mis-
takes, both the object of his science and the limits
of his understanding, who thinks it his business to
explain the means employed by INFINITE WISDOM
for establishing the laws which now govern the
world.

By attending to these obvious considerations, Mr
Kirwan would have avoided a very illiberal and un-
generous proceeding ; and, however he might have
differed from Dr Hutton as to the *truth* of his
opinions, he would not have censured their *tendency* with such rash and unjustifiable severity.

But, if this author may be blamed for wanting
the temper, or neglecting the rules, of philosophic
investigation, he is hardly less culpable, for having
so slightly considered the scope and spirit of a
work which he condemned so freely.   In that
work, instead of finding the world represented as
the result of necessity or chance, which might be
looked for, if the accusations of atheism or impiety were well founded, we see every where the
utmost attention to discover, and the utmost disposition to admire, the instances of wise and benefi-
cent design manifested in the structure, or economy
of the world.   The enlarged views of these, which
his geological system afforded, appeared to Dr
Hutton himself as its most valuable result.   They
were the parts of it which he contemplated with

greatest delight ; and he would have been less flat-
tered, by being told of the ingenuity and originality of his theory, than of the addition which it had
made to our knowledge of *final causes*. It was
natural, therefore, that he should be hurt by an
attempt to accuse him of opinions, so different from
those which he had always taught ; and if he answered Mr Kirwan's attack with warmth or asperity, we must ascribe it to the indignation excited
by unmerited reproach.

120. But to return to the natural history of the
earth : Though there be in it no *data*, from which
the commencement of the present order can be ascertained, there are many by which the existence
of that order may be traced back to an antiquity
extremely remote. The beds of primitive schistus,
for instance, contain sand, gravel, and other materials, collected, as already shown, from the dissolution of mineral bodies ; which bodies, therefore,
must have existed long before the oldest part of the
present land was formed. Again, in this gravel
we sometimes find pieces of sandstone, and of other
compound rocks, by which we are of course carried back a step farther, so as to reach a system of
things, from which the present is the third in succession ; and this may be considered as the most
ancient epocha, of which any memorial exists in
the records of the fossil kingdom.

121. Next in the order of time to the consoli-

dation of the primary strata, we must place their elevation, when, from being horizontal, and at the bottom of the sea, they were broken, set on edge, and raised to the surface. It is even probable, as formerly observed, that to this succeeded a depression of the same strata, and a second elevation, so that they have twice visited the superior, and twice the inferior regions. During the second immersion, were formed, first, the great bodies of pudding-stone, that in so many instances lie immediately above them; and next were deposited the strata that are strictly denominated secondary.

122. The third great event, was the raising up of this compound body of old and new strata from the bottom of the sea, and forming it into the dry land, or the continents, as they now exist. * Contemporary with this, we must suppose the injection of melted matter among the strata, and the consequent formation of the crystallized and unstratified rocks, namely, the granite, metallic veins, and veins of porphyry and whinstone. This, however, is to be considered as embracing a period of great duration; and it must always be recollected, that veins are found of very different formation; so that when we speak generally, it is perhaps impossible to state any thing more precise concerning their antiquity, than that they are posterior to the strata,

---

* NOTE XXI.

and that the veins of whinstone seem to be the most recent of all, as they traverse every other.

123. In the fourth place, with respect to time, we must class the facts that regard the detritus and waste of the land, and must carefully distinguish them from the more ancient phenomena of the mineral kingdom. Here we are to reckon the shaping of all the present inequalities of the surface; the formation of hills of gravel, and of what have been called tertiary strata, consisting of loose and unconsolidated materials; also collections of shells not mineralized, like those in Touraine; such petrifac_ tions as those contained in the rock of Gibraltar, on the coast of Dalmatia, and in the caves of Bayreuth. The bones of land animals found in the soil, such as those of Siberia, or North America, are probably more recent than any of the former. *

124. These phenomena, then, are all so many marks of the lapse of time, among which the principles of geology enable us to distinguish a certain order, so that we know some of them to be more, and others to be less distant, but without being able to ascertain, with any exactness, the proportion of the immense intervals which separate them. These intervals admit of no comparison with the astronomical measures of time; they cannot be expressed by the revolutions of the sun or of the moon; nor

---

* NOTE XXII.

is there any synchronism between the most recent
epochas of the mineral kingdom, and the most ancient of our ordinary chronology.

125. On what is now said is grounded another
objection to Dr Hutton's theory, namely, that the
high antiquity ascribed by it to the earth, is inconsistent with that system of chronology which rests
on the authority of the Sacred Writings.   This objection would no doubt be of weight, if the high
antiquity in question were not restricted merely to
the globe of the earth, but were also extended to
the human race.   That the origin of mankind does
not go back beyond six or seven thousand years, is
a position so involved in the narrative of the Mosaic
books, that any thing inconsistent with it, would no
doubt stand in opposition to the testimony of those
ancient records.   On this subject, however, geology is silent ; and the history of arts and sciences,
when traced as high as any authentic monuments
extend, refers the beginnings of civilization to a
date not very different from that which has just
been mentioned, and infinitely within the limits of
the most recent of the epochas, marked by the physi-
sical revolutions of the globe.

On the other hand, the authority of the Sacred
Books seems to be but little interested in what regards the mere antiquity of the earth itself; nor
does it appear that their language is to be understood literally concerning the *age* of that body, any

more than concerning its *figure* or its *motion.* The
theory of Dr Hutton stands here precisely on the
same footing with the system of Copernicus; for
there is no reason to suppose, that it was the purpose of revelation to furnish a standard of geological,
any more than of astronomical science. It is admitted, on all hands, that the Scriptures are not intended to resolve physical questions, or to explain
matters in no way related to the morality of human
actions ; and if, in consequence of this principle, a
considerable latitude of interpretation were not allowed, we should continue at this moment to believe, that the earth is flat ; that the sun moves
round the earth ; and that the circumference of a
circle is no more than three times its diameter.

It is but reasonable, therefore, that we should
extend to the geologist the same liberty of speculation, which the astronomer and mathematician are
already in possession of ; and this may be done, by
supposing that the chronology of Moses relates only
to the human race. This liberty is not more necessary to Dr Hutton than to other theorists. No
ingenuity has been able to reconcile the natural
history of the globe with the opinion of its recent
origin ; and accordingly the cosmologies of Kirwan
and Deluc, though contrived with more mineralogical skill, are not less forced and unsatisfactory
than those of Burnet and Whiston.

126. It is impossible to look back on the system
which we have thus endeavoured to illustrate, without being struck with the novelty and beauty of the
views which it sets before us. The very plan and
scope of it distinguish it from all other theories of
the earth, and point it out as a work of great and
original invention. The sole object of such theories
has hitherto been, to explain the manner in which
the present laws of the mineral kingdom were first
established, or began to exist, without treating of
the manner in which they now proceed, and by
which their continuance is provided for. The authors of these theories have accordingly gone back
to a state of things altogether unlike the present,
and have confined their reasonings, or their fictions,
to a crisis which never has existed but once, and
which never can return. Dr Hutton, on the other
hand, has guided his investigation by the philosophical maxim, *Causam naturalem et assiduam quæ-
rimus, non raram et fortuitam.* His theory, accordingly, presents us with a system of wise and
provident economy, where the same instruments are
continually employed, and where the decay and renovation of fossils being carried on at the same time
in the different regions allotted to them, preserve
in the earth the conditions essential for the support
of animal and vegetable life. We have been long
accustomed to admire that beautiful contrivance in
nature, by which the water of the ocean, drawn up

in vapour by the atmosphere, imparts, in its descent, fertility to the earth, and becomes the great cause of vegetation and of life ; but now we find, that this vapour not only fertilizes, but creates the soil ; prepares it from the solid rock, and, after employing it in the great operations of the surface, carries it back into the regions where all its mineral characters are renewed. Thus, the circulation of moisture through the air, is a prime mover, not only in the annual succession of the seasons, but in the great geological cycle, by which the waste and reproduction of entire continents is circumscribed. Perhaps a more striking view than this, of the wisdom that presides over nature, was never presented by any philosophical system, nor a greater addition ever made to our knowledge of final causes. It is an addition which gives consistency to the rest, by proving, that equal foresight is exerted in providing for the whole and for the parts, and that no less care is taken to maintain the constitution of the earth, than to preserve the tribes of animals and vegetables which dwell on its surface. In a word, it is the peculiar excellence of this theory, that it ascribes to the phenomena of geology an order similar to that which exists in the provinces of nature with which we are best acquainted ; that it produces seas and continents, not by accident, but by the operation of regular and uniform causes ; that it makes the decay of one part subservient to the restoration of another, and gives stability to the whole, not by perpetuating individuals, but by reproducing them in succession.

127. Again, in the detail of this theory, and the ample induction on which it is founded, we meet with many facts and observations, either entirely new, or hitherto very imperfectly understood. Thus, the veins which proceed from masses of granite, and penetrate the incumbent schistus, had either escaped the observation of former mineralogists, or the importance of the phenomenon had been entirely overlooked. Dr Hutton has described the appearances with great accuracy, and drawn from them the most interesting conclusions. At the junction of the primary and secondary strata, the facts which he has noted had been observed by others; but no one I think had so fully understood the language which they speak, or had so clearly perceived the consequences that necessarily follow from them. He is the first who distinctly pointed out the characters which distinguish whinstone from lava, and who explained the true relation that subsists between these substances. He also discovered the induration of the strata, in contact with veins of whin, and the charring of the coal in their vicinity. His theory also enabled him to determine the affinity of whinstone and granite to one another, and their relation to the other great bodies of the mineral kingdom.

ɪ

To the observations of the same excellent geologist, we are indebted for the knowledge of the general and important fact, that all the hard substances of the mineral kingdom, when elevated into
the atmosphere, have a tendency to decay, and are
subject to a disintegration and waste, to which no
limit can be set but that of their entire destruction ; that no provision is made on the surface for
repairing this waste, and that there, no new fossil
is produced ; that the formation of all the varied
scenery which the surface of the earth exhibits, depends on the operation of causes, the momentary
exertions of which are familiar to us, though we
knew not before the effects which their accumulated action was able to produce.   These are facts in
the natural history of the earth, the discovery of
which is due to Dr Hutton ; and, should we lay all
further speculation aside, and consider the theory
of the earth as a work too great to be attempted by
man, we must still regard the phenomena and laws
just mentioned, as forming a solid and valuable addition to our knowledge.

128. If we would compare this theory with
others, as to the invisible agents which it employs,
we must consider, that fire and water are the two
powers which all of them must make use of, so that
they can differ from one another only by the way
in which they combine these powers.   In Dr Hutton's system, water is first employed to deposit and

arrange, and then fire to consolidate, mineralize, and lastly, to elevate the strata ; but, with respect to the unstratified or crystallized substances, the action of fire only is recognised.   The system having least affinity to this is the Neptunian, which ascribes the formation of all minerals to the action of water alone, and extends this hypothesis even to the unstratified rocks.   Here, therefore, the action of fire is entirely excluded ; and the Neptunists have certainly made a great sacrifice to the love of truth, or of paradox, in rejecting the assistance of so powerful an auxiliary. *

129. In the systems which employ the agency of the latter element, we are to look for a greater resemblance to that of Dr Hutton, though many and great marks of distinction are easily perceived. In the cosmologies, for example, of Leibnitz and Buffon, fire and water are both employed, as well as in this ; but they are employed in a reverse order.   These philosophers introduce the action of fire first, and then the action of water, which is to invert the order of nature altogether, as the consolidation of the rocks must be posterior to their stratification.   Indeed, the theory of Buffon is singularly defective : besides inverting the order of the two great operations of stratification and consolidation, and of course giving no real explanation of

---

* NOTE XXIII.

the latter, it gives no account of the elevation, or highly inclined position of the strata ; it makes no distinction between stratified and unstratified bodies, nor does it offer any but the most unsatisfactory explanation of the inequalities of the earth's surface. This system, therefore, has but a very distant resemblance to the Huttonian theory. *

130. The system of Lazzaro Moro has been remarked as approaching nearer to this theory than any other ; and it is certain, that one very important principle is common to them both. The theory of the Italian geologist was chiefly directed to the explanation of the remains of marine animals, which are found in mountains far from the sea ; and it appears to have been suggested to him by the phenomena of the *Campi Phlegræi,* and by the production of the new island of *Santorini* in the Archipelago. He accordingly supposes, that the islands and continents have been all raised up, like the above-mentioned island, from the bottom of the sea, by the force of volcanic fire : that these fires began to burn under the bottom of the ocean, soon after the creation of the world, when as yet the ocean covered the whole earth : that they at first elevated a portion of the land ; and in this primitive land no shells are found, as the original ocean was destitute of fish. The volcanoes continuing to

---

* NOTE XXIV.

burn, under the sea, after the creation of animated nature, the strata that were then raised up by their action were full of shells and other marine objects; and, from the violence with which they were elevated, arose the coutortions and inclined position which they frequently possess. *

This system is imperfect, as it makes no peculiar provision for the consolidation of the strata, which, according to it, as well as the Neptunian system, must be ascribed to the action, not of fire, but of water. No account is given of the mineralization of the shells found in the strata, or of the difference between them and the shells found loose at the bottom of the sea; and no distinction is made between stratified and unstratified substances. But, with all this, Lazzaro Moro has certainly the merit of having perceived, that some other power than that which deposited the strata, must have been employed for their elevation, and that they have endured the action of a disturbing force.

131. From this comparison it appears, that Dr Hutton's theory is sufficiently distinct, even from the theories which approach to it most nearly, to merit, in the strictest sense, the appellation of *new* and *original.* There are indeed few inventions or discoveries, recorded in the history of science, to

---

* Dé Crostacei, e degli altri Marini Corpi, che si trovano su' Monti: di Ant. Lazzaro Moro. Venezia. 1740.

which nearer approaches were not made before they were fully unfolded. It therefore very well deserves to be distinguished by a particular name ; and, if it behoves us to follow the analogy observed in the names of the two great systems, which at present divide the opinions of geologists, we may join Mr Kirwan in calling this the PLUTONIC SYSTEM. For my own part, I would rather have it characterized by a less splendid, but juster name, that of the HUTTONIAN THEORY.

132. The circumstance, however, which gives to this theory its peculiar character, and exalts it infinitely above all others, is the introduction of the principle of pressure, to modify the effects of heat when applied at the bottom of the sea. This is in fact the key to the grand enigma of the mineral kingdom, where, while one set of phenomena indicates the action of fire, another set, equally remarkable, seems to exclude the possibility of that action, by presenting us with mineral substances, in such a state as they could never have been brought into by the operation of the fires we see at the surface of the earth. These two classes of phenomena are reconciled together, by admitting the power of compression to confine the volatile parts of bodies when heat is applied to them, and to force them, in many instances, to undergo fusion, instead of being calcined or dissipated by burning or inflammation. In this hypothesis, which

some affect to consider as a principle gratuitously
assumed, there appears to me nothing but a very
fair and legitimate generalization of the properties
of heat. Combustion and inflammation are chemical processes, to which other conditions are required, besides the presence of a high temperature.
The state of the mineral regions makes it reasonable to presume, that these conditions are wanting
in the bowels of the earth, where, of consequence,
we have a right to look for nothing but expansion
and fusion, the only operations which seem essential to heat, and inseparable from the application
of it, in certain degrees, to certain substances.
Though this principle, therefore, had no countenance from analogy, the admirable simplicity, and
the unity, which it introduces into the phenomena
of geology, would sufficiently justify the application of it to the theory of the earth.

As another excellence of this theory, I may,
perhaps, be allowed to remark, that it extends its
consequences beyond those to which the author of
it has himself adverted, and that it affords, which
no geological theory has yet done, a satisfactory explanation of the spheroidal figure of the earth. *

133. Yet, with all these circumstances of originality, grandeur, and simplicity in its favour, with
the addition of evidence as demonstrative as the

* NOTE XXV.

nature of the subject will admit, this theory has probably many obstacles to overcome, before it meet the general approbation. The greatness of the objects which it sets before us, alarms the imagination ; the powers which it supposes to be lodged in the subterraneous regions ; a heat which has subdued the most refractory rocks, and has melted beds of marble and quartz ; an expansive force, which has folded up, or broken the strata, and raised whole continents from the bottom of the sea ; these are things with which, however certainly they may be proved, the mind cannot soon be familiarized. The change and movement also, which this theory ascribes to all that the senses declare to be most unalterable, raise up against it the same prejudices which formerly opposed the belief in the true system of the world ; and it affords a curious proof, how little such prejudices are subject to vary, that as Aristarchus, an ancient follower of that system, was charged with impiety for moving the everlasting VESTA from her place, so Dr Hutton, nearly on the same ground, has been subjected to the very same accusation. Even the length of time which this theory regards as necessary to the revolutions of the globe, is looked on as belonging to the marvellous ; and man, who finds himself constrained by the want of time, or of space, in almost all his undertakings, forgets, that in these, if in any thing, the riches of nature reject all limitation. *

The evidence which must be opposed to all these causes of incredulity, cannot be fully understood without much study and attention. It requires not only a careful examination of particular instances, but comprehensive views of the whole phenomena of geology; the comparison of things very remote with one another; the interpretation of the *obscure* by the *luminous,* and of the *doubtful* by the *decisive* appearances. The geologist must not content himself with examining the insulated specimens of his cabinet, or with pursuing the nice subtleties of mineralogical arrangement; he must study the relations of fossils, as they actually exist; he must follow nature into her wildest and most inaccessible abodes; and must select, for the places of his observations, those points, from which the variety and gradation of her works can be most extensively and accurately explored. Without such an exact and comprehensive survey, his mind will hardly be prepared to relish the true theory of the earth. " *Naturæ enim vis atque majestas omnibus momentis fide caret, si quis modo partes atque non totam complectatur animo.*" †

134. If indeed this theory of the earth is as

---

* Note xxvi.
† Plin. Hist. Nat. Lib. VII. Cap. i.

well founded as we suppose it to be, the lapse of
time must necessarily remove all objections to it,
and the progress of science will only develope its
evidence more fully. As it stands at present,
though true, it must be still imperfect; and it cannot be doubted, that the great principles of it,
though established on an immoveable basis, must
yet undergo many modifications, requiring to be limited, in one place, or to be extended, in another.
A work of such variety and extent cannot be carried to perfection by the efforts of an individual.
Ages may be required to fill up the bold outline
which Dr Hutton has traced with so masterly a
hand; to detach the parts more completely from
the general mass; to adjust the size and position
of the subordinate members; and to give to the
whole piece the exact proportion and true colouring of nature.

This, however, in length of time, may be expected from the advancement of science, and from
the mutual assistance which parts of knowledge,
seemingly the most remote, often afford to one
another. Not only may the observations of the
mineralogist, in tracts yet unexplored, complete
the enumeration of geological facts; and the experiments of the chemist, on substances not yet subjected to his analysis, afford a more intimate acquaintance with the nature of fossils, and a measure
of the power of those chemical agents to which

this theory ascribes such vast effects; but also, from other sciences, less directly connected with the natural history of the earth, much information may be received. The accurate geographical maps and surveys which are now making; the soundings; the observations of currents; the barometrical measurements, may all combine to ascertain the reality, and to fix the quantity of those changes which terrestrial bodies continually undergo. Every new improvement in science affords the means of delineating more accurately the face of nature as it *now* exists, and of transmitting, to future ages, an account, which may be compared with the face of nature as it shall *then* exist. If, therefore, the science of the present times is destined to survive the physical revolutions of the globe, the HUT- TONIAN THEORY may be confirmed by historical record; and the author of it will be remembered among the illustrious few, whose systems have been verified by the observations of succeeding ages, supported by facts unknown to themselves, and established by the decisions of a tribunal, slow, but infallible, in distinguishing between truth and falsehood.

NOTES AND ADDITIONS.

NOTES AND ADDITIONS.

_____

NOTE I. § 2.

Origin of Calcareous Rocks.

135. It has been asserted, that Dr Hutton went farther than is stated at § 2, and maintained all calcareous matter to be *originally* of animal formation. This position, however, is so far from being laid down by Dr Hutton, that it belongs to an inquiry which he carefully avoided to enter on, as being altogether beyond the limits of philosophical investigation.

He has indeed no where treated of the *first origin* of any of the earths, or of any substance whatsoever, but only of the transformations which bodies have undergone since the present laws of nature were established. He considered this last as all that a science, built on experiment and observation, can possibly extend to ; and willingly left, to more presumptuous inquirers, the task of carrying their reasonings beyond the boundaries of nature, and of unfolding the properties of the chaotic fluid, with as much minuteness of detail, as if

they were describing the circumstances of a chemical process which they had actually witnessed.

The idea of calcareous matter which really belongs to the Huttonian Theory, is, that in all the
changes which the terraqueous globe has undergone in past ages, this matter existed, as it does
now, either in the form of limestone and marble,
or in the composition of other stones, or in the
state of corals, shells, and bones of animals. It
may be true, that there is no particle of calcareous
matter, at present existing on the surface of the
earth, that has not, at some time, made a part of
an animal body; but of this we can have no certainty, nor is it of any importance that we should.
It is enough to know, that the rocks of marble
and limestone contain in general marks of having
been formed from materials collected at the bottom of the sea; and of this a single cockle-shell, or
piece of coral, found included in a rock, is a sufficient proof with respect to the whole mass of which
it makes a part.

The principal object which Dr Hutton had in
view when he spoke of the masses of marble and
limestone, as composed of the calcareous matter
of marine bodies, * was to prove, that they had
been all formed at the bottom of the sea, and from
materials there deposited. His general conclusion

---

* Theory of the Earth, Vol. I. p. 23, 24.

is, " That all the strata of the earth, not only those consisting of such calcareous masses, but others superincumbent upon these, have had their origin at the bottom of the sea, by the collection of sand and gravel, of shells, of coralline and crustaceous bodies, and of earths and clays variously mixed, or separated and accumulated. This is a general conclusion, well authenticated by the appearances of nature, and highly important in the natural history of the earth." *

136. In his Geological Essays, Mr Kirwan says, that " some geologists, as Buffon, and of late Dr Hutton, have excluded calcareous earth from the number of the primeval, asserting the masses of it we at present behold to proceed from shell-fish. But, in addition to the unfounded supposition, that shell-fish, or any animals, possess the power of producing any simple earth, these philosophers should have considered, that, before the existence of any fish, the stony masses that inclose the bason of the sea, must have existed ; and, among these, there is none in which calcareous earth is not found. Dr Hutton endeavours to *evade* this argument, by supposing the world we now inhabit to have arisen from the ruins and fragments of an anterior, without pointing at any original. If we are thus to pro-

---

* Theory of the Earth, Vol. I. p. 26.

ceed *in infinitum,* I shall not pretend to follow him ;
but, if he stops any where, he will find the same
argument equally to occur." *

The argument here employed would certainly be
conclusive against any one, who, in disputing about
the *first origin* of things, should deny that the calcareous is as ancient as any other of the simple
earths.   But this has nothing to do with Dr Hutton's speculations, which, as has been just said, never
extended to the *first origin* of substances, but were
confined entirely to their changes ; so that what he
asserts concerning the calcareous rocks, is no more
than that those which we now see have been formed from loose materials, deposited at the bottom of
the sea.   It was not therefore in order to *evade* Mr
Kirwan's argument, as the preceding passage would
lead us to believe, that he supposed the world which
we now inhabit to have arisen from the ruin and
waste of an anterior world ; but it was because this
seemed to him a conclusion which necessarily followed from the phenomena of geology, and it was a
conclusion that he had deduced long before he heard
of Mr Kirwan's objections to his system.   Instead
of an *evasion,* therefore, any one who considers the
subject fairly, will see, in Dr Hutton's reasoning,
nothing but the caution of a philosopher, who wise-

* Geol. Essays, p. 13.

11

ly confines his theory within the same limits by which nature has confined his experience and observation.

It is nevertheless true, that Dr Hutton has sometimes expressed himself as if he thought that the present calcareous rocks are all composed of animal remains. * This conclusion, however, is more general than the facts warrant ; and, from some incorrectness or ambiguity of language, is certainly more general than he intended. The idea of calcareous rocks, on which he argues throughout his whole theory, is precisely that which is stated in the preceding article.

NOTE II. § 6.

Origin of Coal.

137. The vegetable origin of coal seems to be sufficiently proved by the reasoning in § 5 and 6 ; and that reasoning will appear still more satisfactory, from what is said at § 28 and 29, concerning the consolidation of this fossil. Dr Hutton has treated both of the matter of coal and of its consolidation, Part I. Chap. 8, of his Theory of the Earth. †

---

* Theory of the Earth, Vol. I. p. 23.

† Vol. I. p. 558, &c.

The notion, however, that coal is of vegetable origin, is not peculiar to this theory, but has been for some time the prevailing opinion. Buffon supposes this mineral to be formed from vegetable and animal substances, the oil and fat of which have been converted into bitumen by the action of acids. * A fundamental mistake, however, is committed by this author, and by M. Gensanne, (author of the natural history of Languedoc,) on whose observations he greatly relies, in considering coal as consisting of bitumen united to earth, thus omitting the only ingredient essential to coal, namely the carbon or charcoal. This may truly be considered as the essential part, because coal may exist without bitumen, as in the instance of blind coal, but not without charcoal.

Another theory of coal, very analogous to Dr Hutton's, is that of Arduino, professor of mineralogy at Venice, in which he supposes it formed from vegetable and animal remains from the land and sea, but chiefly from the latter. † This theory of coal is contained in Dr Hutton's, in which the animal and vegetable remains must be supposed to come both from the earth and the sea. It seems to be without any good reason that Arduino considers

---

* Hist. Nat. des Mineraux, Tom. I. p. 429, 4to edit.

† Saggio Fisico-mineralogico del Sig. Giov. Arduino; Atti di Siena, Tom. V. p. 228, 281, &c.

10

the sea as the chief source of these materials. His remarks, however, are very ingenious, and deserving of attention.

These accounts of the origin of coal are all nearly the same ; it is in what relates to the distinction between the common coal, in which there is no ligneous structure, and those varieties of it in which that structure is apparent, and again in explaining the consolidation of both, that the theory laid down here is peculiar.

138. Some other mineralogists refer one of the ingredients of coal to the vegetable kingdom, but not the other. Unable to resist the conviction which arises from the fibrous structure of parts of strata, and even entire strata of coal, they have supposed, that wood, which had been somehow buried in the earth, or perhaps deposited at the bottom of the sea, had become impregnated with bitumen. which last, however, they consider as of mineral origin. This appears to be the opinion of Lehman, and also of some very late writers. There seems, however, to be hardly less reason for referring the origin of one part of coal to the vegetable or animal kingdom than another. The two last are certainly capable of furnishing both the carbonic and bituminous parts ; and therefore, to derive these from different sources, is at least a very unnecessary complication of hypotheses.

139. Another explanation of coal, very dif-

ferent from any of the preceding, has lately been
advanced and set up in opposition to the Huttonian Theory. Mr Kirwan, * the only mineralogist, I believe, who has attempted to derive both
the carbonic and bituminous matter of coal from
the mineral kingdom, distinguishes between wood
coal and mineral coal, and gives a theory entirely
new of the formation of the latter. Wood coal
is that in which the ligneous structure is so apparent, as to leave no doubt of its vegetable origin ;
mineral coal is that in which no such structure can
be discovered, and is the same which Dr Hutton
derives from the vegetable juices, and other remains, comminuted, dispersed, carried into the sea,
and there precipitated, so as to unite with different
proportions of earth, and to become afterwards
mineralized.

These two species of coal, which the Huttonian
theory considers as gradations of the same substance, Mr Kirwan regards as perfectly distinct,
constituting two minerals, of an origin and formation entirely different. He therefore endeavours
to ascertain the distinguishing characters of each,
considered geologically.

140. But here the leading distinction, implied
in all the rest, that the two kinds of coal are never
found in the same bed, but always in different

---

* Geol. Essays, Essay vii. p. 290.

situations, and with different laws of stratification, is expressly contradicted by matter of fact. Coal, as is said above, with its ligneous texture quite apparent, and coal with no such structure visible, are often found in the same seam, are brought up from the same mine, and united in the same specimen. I have a specimen from a bed of coal, in the Isle of Sky, found under a basaltic rock, consisting of a ligneous part, which graduates into one in which there is no vestige of a fibrous texture, and in which the surface is smooth and glossy, with a fracture almost vitreous. The upper part of the specimen is therefore perfect wood coal, and the under part perfect mineral coal, in the language of Mr Kirwan ; at the same time that the transition from the one to the other is made by insensible degrees. This specimen, were it perfectly solitary, is sufficient to prove the identity of the two species of coal we are now speaking of, and to show, that the difference between them is accidental, not essential. The specimen, however, is far from being solitary ; the number of similar appearances is so great, as hardly to have escaped the observation of any mineralogist. Mr Kirwan admits, that wood coal is often found under basaltes ; * but what is essential to be remarked is, that, in this instance, we have both the wood coal and

* Geol. Essays, p. 310.

the common mineral coal, lying under that rock, and the one passing gradually into the other. It appears, indeed, that many of the facts which Mr Kirwan produces, in treating of what he calls *carboniferous* soils, are quite inconsistent with the distinction he would make between wood-coal and mineral coal. *

141. It is, however, true, that there are instances in which the wood coal, or fossil wood, as it is usually called, forms entire beds, quite unconnected with the ordinary coal, and stratified in some respects differently. Such is the Bovey coal in Devonshire, the wood coal in the north of Ireland, and perhaps the Surturbrandt of Iceland. With respect to the Bovey coal, it does by no means answer to one of Mr Kirwan's remarks, viz. that late observations have ascertained, that no such parallelism of the beds, as in mineral coal, nor even any distinct number of strata, is found. In the Bovey coal, the number of strata is very well defined, by beds of clay regularly interposed ; but as to the extent of these beds, the coal having been worked only at one place, and by an open pit, without any extensive subterraneous excavation, nothing is known with certainty.

In the Bovey coal too, I must observe, though its beds have the ligneous structure very distinct,

* Geol. Essays, p. 311.

the clay interposed between these beds, which is but little indurated, contains a great deal of coaly matter, in the form of thin flakes, interspersed through it. So far as I know, there are no mineral veins nor shifts, nor any bed of indurated stone, that accompany this coal; so that, though one cannot doubt of its vegetable origin, some doubt may be entertained concerning the nature of the mine- ralizing operations, to which it has been subjected. The consideration of these, however, does not belong to the present argument; and the peculiarities of this semi-mineralized coal, as it may be called, have nothing to do with the general question, whether wood coal and mineral coal are the same substance; about which question, if the gradations are properly considered, I think, no reasonable doubt can remain.

142. One of Mr Kirwan's objections to the vegetable origin of coal, is founded on this fact, that there is, in the museum at Florence, a cellular sandstone, the cells of which are filled with genuine mineral coal. " Could this (adds he) have been originally wood ?" * The answer to the interrogatory proposed here as a *reductio ad absurdum*, is, that most undoubtedly it may have been wood. Sandstone with charred wood, that is, with wood- coal in it, is not an uncommon phenomenon in coal

---

* Geol. Essays, p. 321.

countries. I have seen a specimen of this kind
from the Hales Quarry, near Edinburgh, consisting
of a piece of charred wood, imbedded in sandstone;
the wood was much altered, but the remains of its
fibrous structure were distinctly visible.   This affords a perfect commentary on the specimen in the
Florence cabinet.

143. If then it be granted, as I think it must,
that the two kinds of coal we have been speaking of
are of the same origin, it is not very necessary to
enter on a refutation of Mr Kirwan's theory with
respect to either of them.   His account of the formation of mineral coal, however, is so singular, that
it cannot be passed over without remark.

Mr Kirwan supposes, 1mo, That natural carbon
was originally contained in many mountains of the
granite and porphyritic order, and also in siliceous
schistus ; and might, by disintegration and decomposition, be separated from the stony particles. 2do,
That both petrol and carbon are often contained in
trap, since hornblende, which has lately been found
to contain carbon, very frequently enters into its
composition.

" My opinion (adds he) is, that coal mines, or
strata of coal, as well as the mountains in which
they are found, owe their origin to the disintegration of primeval mountains, either now totally destroyed, or whose height and bulk, in consequence of
such disintegration, are considerably lessened ; and

that these rocks, anciently destroyed, contained most probably a far larger proportion of carbon and petrol than those of the same denomination now contain, since their disintegration took place at so early a period. *

" By the decomposition of these mountains, the feldspar and hornblende were converted into clay ; the bituminous particles, thus set free, reunited, and were absorbed, partly by the argil, but chiefly by the carbonaceous matter, with which they have the greatest affinity. The carbonic and bituminous particles, thus united, being difficultly miscible with water, and specifically heavier, sunk through the moist, pulpy, incoherent argillaceous masses, and formed the lowest stratum," &c.

Such is Mr Kirwan's theory of the formation of coal, and nobody I think will dispute the originality of it.

144. To enter on a formal refutation of an opinion so loaded with objections, would be a task as irksome as unnecessary.   A few observations will suffice.

The notion of the great degradation of mountains, involved in this hypothesis, is the part of it to which I am least disposed to object.   But I cannot help reminding Mr Kirwan, that the effects of waste are not supposed less in this, than in Dr

----

* Geol. Essays, p. 328, &c.

Hutton's theory; and that he has assumed the
very principle, of which that theory makes so much
use, though he has reserved to himself, as it should
seem, the right of denying it, when it does not accord with his system.   It is indeed worth while to
compare what is said concerning the degradation of
mountains, in the above quotations, and still more
fully in the book itself, with what is advanced concerning their indestructibility, in another passage
of the same volume : *

  " All mountains are not subject to decay ; for
instance, scarce any of those that consist of red
granite.   The stone of which the Runic rocks are
formed, have withstood decomposition for two
thousand years, as their characters evince," &c.

  " Basaltic pillars, in general, bid defiance to decay," &c.   He goes on to deny every step of the
degradation of land, by which it is wasted, carried
into the sea, and spread out over its bottom, though
all these are necessary *postulata* in his theory of
the formation of coal.   One can be at no loss about
estimating the value of a system, in which such
gross inconsistencies make a necessary part.

  145. The quantity of hornblende and siliceous
schistus, necessary to be decomposed, in order to
produce the coal strata presently existing, is enormous, and would lead to an estimate of what is

* Page 436.

worn away from the primeval mountains, far exceeding any thing that Dr Hutton has supposed.
It is true, that Mr Kirwan, never at all embarrassed
about preserving a similitude between nature as she is
now, and as she was heretofore, lays it down, that the
part of the primeval mountains which is worn away,
contained much more carbon than the part which
is left behind. This, however, is an arbitrary supposition ; and since, in this system, such suppositions are so easily admitted, why may we not conceive, in the primeval mountains, a more copious
source of carbonic matter than hornblende or siliceous schistus ? We have but to imagine, that the
diamond existed among these mountains in such
abundance, as to constitute large rocks. This stone
being made up of pure, or highly concentrated carbon, the adamantine summits of a single ridge, by
their decomposition, might afford a carbonic basis,
sufficient for the coal beds of all the surrounding
plains.

146. We may also object to Mr Kirwan, that
the siliceous part of the mountains has not been
chemically dissolved ; it has been only abraded
and worn away. Mechanical action has reduced
the quartz to gravel and sand, but has not produced on it any chemical change. The carbon, therefore, could not be let loose. Experiment, indeed,
might be employed, to determine whether the siliceous matter of the secondary, and of the primary

strata contains this substance in the same proportion.

Again, a more fatal symptom can hardly be imagined in any theory, than that, when the circumstances of the phenomena to be explained are *a little* changed, the theory is under the necessity of
changing *a great deal.* Now, this is what happens to Mr Kirwan's theory, in the attempt made
to explain by it the stratum of coal described in
the *Annales de Chimie,* * as cutting a mountain of
argillaceous strata in two, at about three-fourths of
its height. This stratum, Mr Kirwan says, must
have been formed by *transudation* from the superior part of the mountain. † Besides that this is a
gratuitous supposition of a thing, without example,
it involves in it an absurdity, which becomes evident the moment the question is asked, What occupied the place of the coal-bed before the transudation from the upper part of the mountain ? Has
the *liquid coal,* as it percolated through the upper
strata, expelled any substance from the place it now
occupies ? or has it been powerful enough to raise
up, or to float, as it were, the upper part of the
mountain ?

The situation of this bed of coal is not singular,
and its formation is easily explained on Dr Hutton's theory. It is part of a stratum of coal,

* Tom. XI. p. 272.   † Geol. Essays, p. 338.

which has been deposited, like all others, at the
bottom of the sea ; from whence certain causes, of
very general operation, have raised it up, together
with the attending strata : these strata have since
been all cut down, and worn away by the operations of the surface ; and the mountain, with the
coal stratum in the middle of it, is a part of them
which has been left behind.   There is no wonder,
that a coal stratum should be found alternating
with others, in a mountain, any more than in the
bowels of the earth, and no more need of a separate
explanation. *

147. After all, it may be asked, for what purpose is it that so many incongruous and ill supported hypotheses are thus piled on one another ? is it
only to avoid ascribing the carbonic and bituminous matter of coal to a substance in which we
know with certainty that such matter resides in
great abundance, in order to derive it from other
substances, in which a subtle analysis has shown,
that it exists in a very small proportion ?   Such
reasoning is so great a trespass on every principle
of common sense, not to say of sound philosophy,
that, to bestow any time on the refutation of it, is,
in some degree, to fall under the same censure.

---

* This stratum of coal, which is described by Hassenfratz,
is remarkable for being in a mountain which rests immediately on primary schistus and granite.

NOTE III. § 7.

Primitive Mountains.

148. The enumeration of the different kinds of primary schistus, at § 7, is not proposed as at all complete. It will be less defective, however, if we add to it *talcose schistus,* and *lapis ollaris* or *potstone.* *

149. The rocks called here by the name of primary, were first distinguished, as forming the basis of all the great chains of mountains, and as constituting a separate division of the mineral kingdom, by J. G. Lehman, director of the Prussian mines. See his work, entituled, Essai d'une Histoire Naturelle des Couches de la Terre. † These rocks were regarded by Lehman as parts of the original *nucleus* of the globe, which had undergone no alteration, but remained now such as they were at first created ; and, agreeably to this supposition, he bestowed on them, and on the mountains composed of them, the name of primitive. He remarks, nevertheless, their distribution into beds, either perpendicular to the horizon, or highly in-clined, and the super-position of the secondary

---

* Kirwan's Mineralogy, Vol. I. p. 155.

† Tom. III. p. 239, &c. The French translation is in 1759, but the original preface is dated at Berlin, 1756.

and horizontal strata. However mineralogists may now differ in their theories from Lehman, they must consider this distinction as a great step in the science of geology, and very material to the right arrangement of the natural history of the earth.

150. Several mineralogists have agreed with him in the supposition, that these rocks are a part of the original structure of the globe, and prior to all organized matter. Of this number is Pallas; * and also Deluc, who applies the term *primordial* to the rocks in question, and considers them as neither stratified nor formed by water. † In his subsequent writings, however, he admits their formation from aqueous deposition, as the Neptunists do in general, but holds them to be more ancient than organized bodies.

151. Pini, professor of natural history at Milan, has denied the stratification of primitive mountains, in a memoir on the mineralogy of St Gothard, and in another on the revolutions of the globe. ‡ His reasonings are opposed by Saussure, §

---

* Observations sur la Formation des Montagnes.

† Lettres Phys. sur l'Histoire de la Terre, Tom. II. p. 206.

‡ Memoria sulle Rivoluzioni del Globo Terrestre; Memorie della Società Italiana, Tom. V. p. 222, &c.

§ Voyages aux Alpes, Tom. IV. § 1881.

and are certainly, in many respects, very open to attack. They proceed on a comparison between the division of rocks, by what is called the planes of their stratification, and their division by transverse fissures : two things, which he thinks so much alike, that they ought not to be referred to different causes ; and, as the one cannot be regarded as the effect of aqueous deposition, so neither should the other. This is a very fallacious argument, because it confounds two things that are essentially different ; and, instead of inquiring about a matter of fact, inquires about its cause. The truth is, that the dispute has arisen from not distinguishing the granite from the schistus mountains, and from involving both under the name of primitive. M. Pini seems to be in the right, when he holds the granite of St Gothard to be unstratified ; but it is without any good reason, that he would extend the same conclusion to the schistus of that mountain. Charpentier, and Saussure, in his last two volumes, contend even for the stratification of granite. *

As the consent, if not universal, is very general for the stratification of the primary schistus, and the fact itself abundantly obvious, in almost all the instances I have ever met with, I have not

---

* See Note xv. on Granite.

11

considered it as necessary to enter here into any argument on this subject.

Note iv. § 8.

Primary Strata not Primitive.

152. An account of the facts referred to § 8, may be found in Hutton's Theory, Vol. I. p. 332, &c. To what is there said, of the shells contained in the primary limestone of Cumberland, I must add, that I have since had an opportunity of verifying the conjecture, that the limestone rock, in which the shells were found, near the head of Coniston Lake, is part of the same body of strata, where shells were found, in a quarry between Ambleside and Low-wood. The limestone of that quarry contains several marine objects; it is in strata declining about 10° from the perpendicular, toward the S. E., and forms a belt, stretching across the country from N. E. to S. W.

In a quarry where the argillaceous schistus, on the south side of this limestone belt, is worked for pavement, are impressions of what I think may safely be accounted marine objects; they have the form of shells, are much indurated, and full of pyrites. They seem to be of the same kind with the impressions said to be found in a slate quarry, near the village of Mat in Switzerland. *

Another spot, affording instances of shells in primary limestone, is in Devonshire.   On the sea shore on the east side of Plymouth Dock, opposite to Stonehouse, I found a specimen of schistose micaceous limestone, containing a shell of the bivalve kind : it was struck off from the solid rock, and cannot possibly be considered as an adventitious fossil.

Now, no rocks can be more decidedly primary than those about Plymouth.   They consist of calcareous strata, in the form either of marble or micaceous limestone, alternating with varieties of the same schistus, which prevails through Cornwall to the west, and extends eastward into Dartmoor, and on the sea coast, as far as the Berry-head.   These all intersect the horizontal plane, in a line from east to west nearly ; they are very erect, those at Plymouth being elevated to the north.

Though, therefore, the remains of marine animals are not frequent among the primary rocks, they are not excluded from them ; and hence the existence of shell-fish and zoophytes, is clearly proved to be anterior to the formation even of those parts of the present land which are justly accounted the most ancient.

---

* Hutton's Theory, Vol. I. p. 327.

153. The rocks which contain sand or gravel, or which are of a granulated texture, must also be considered as carrying in themselves a testimony of the most unequivocal kind, of their being derived from the *detritus* and waste of former rocks. Now, the fact stated in the text, concerning sand found in schistus, most justly accounted primary, might be exemplified by actual reference to many spots on the earth's surface. A few such will be sufficient in this place.

St Gothard is a central point, in one of the greatest tracts of primary mountains on the face of the earth, yet arenaceous strata are found in its vicinity. Between Airolo and the Hospice of St Gothard, Saussure found a rock, composed of an arenaceous or granular paste, including in it hornblende and garnets. He is somewhat unwilling to give the name *grés* to this stone, which M. Besson had done ; but he nevertheless describes it as having a granulated structure. *

Among the most indurated rocks that compose the mountains of this island, many are arenaceous. Thus, on the western coast of Scotland, the great body of high and rugged mountains on the shores of Arasaig, &c. from Ardnamurchan to Glenelg, consists, in a great measure, of a granitic sandstone, in vertical beds. This stone sometimes occupies

---

* Voyages aux Alpes, Tom. IV. § 1822.

great tracts; at other times it is alternated with
the micaceous, or other varieties of primary schistus; it occurs, likewise, in several of the islands,
and is a fossil which we hardly find described or
named by the writers on mineralogy.   Much, also,
of a highly indurated, but granulated quartz, is
found in several places in Scotland, in beds or
strata, alternated with the common schistus of the
mountains.   Remarkable instances of this may be
seen on the north side of the ferry of Balachulish,
and again on the sea shore at Cullen.   At the latter, the strata are remarkably regular, alternating
with different species of schistus.   At the former,
the quartz is so pure, that the stone has been mistaken for marble.

These examples are perhaps sufficient; but I
must add, that in the micaceous and talcose schisti
themselves, thin layers of sand are often found, interposed between the layers of mica or talc.   I
have a specimen, from the summit of one of the
highest of the Grampian mountains, where the thin
plates, of a talcky or asbestine substance, are separated by layers of a very fine quartzy sand, not
much consolidated.

The mountain from which it was brought, consists of vertical strata, much intersected by quartz
veins.   It is impossible to doubt, in this instance,
that the thin plates of the one substance, and the
small grains of the other, were deposited together

at the bottom of the sea, and that they were alike produced from the degradation of rocks, more ancient than any which now exist.

154. In the Neptunian system, as improved by Werner, an attempt is made to take off the force of such instances as are produced in § 8, 9, and 152, &c. by distinguishing rocks, as to their formation, into three different orders, the primitive; the intermediate, and the secondary, or, to speak more properly, into primary, secondary, and tertiary. The same mineralogist distinguishes, among the materials of these rocks, between what he terms chemical and mechanical deposits. By mechanical deposits, are understood sand, gravel, and whatever bears the mark of fracture and attrition ; by chemical deposits, those which are regularly crystallized, or which have a tendency to crystallization, and in which the action of mechanical causes cannot be traced. This distinction is founded in nature, and proceeds on real and palpable differences ; but the application made of it to the three kinds of strata just enumerated, seems by no means entitled to the same praise.

The primitive rocks contain, it is said, none but chemical deposits, and are entirely composed of them : the intermediate contain a mixture of both, and also some vestiges of organized bodies : the secondary consist almost entirely of the mechanical, or of the remains of such bodies, with little of the chemical. The first of these, then, are held to contain no mark or vestige whatsoever of any thing more ancient than themselves, and are, in the strictest sense, primeval, or formed of the first materials, deposited by the immense ocean which originally encompassed the globe.

After them were formed the intermediate, mostly consisting of chemical deposits, but containing also some animal remains, and some spoils from the land, subjected to the various kinds of destruction, which even then made a part of the order of nature. These rocks, it is alleged, are chiefly argillaccous, are less indurated than the primary, and not intersected by veins of quartz.

The secondary were formed from the remains of the other two, and contain more mechanical deposits than any other.

This sketch of what I understand to be Werner's opinion concerning the different formation of the strata, is chiefly taken from a view of his system, in the *Journal de Physique* for 1800.

155. The main objection to the distinction here made between the primary and the intermediate strata, is founded on the facts that have been just stated. The sandstone of St Gothard is from a country having every character of a primary one in the highest perfection. The instances I have mentioned from the Highlands of Scotland, are from mountains, less elevated indeed than the Alps, but

where the rock is micaceous, talcose, or siliceous, in planes erect to the horizon, and intersected by veins of quartz. The shells from Plymouth are from a rock, that Werner would, I think, admit to be truly primitive. Those from the lakes, also, are from the centre of a country, occupied by porphyry, schorl, hornstone-schistus, and many others, about the order of which there can be no dispute. It is true, that in this tract there are argillaceous strata, of the kind that might be accounted intermediate, were they not interposed among those that are certainly primary; and this very intermixture shows, how little foundation there is for the distinction attempted to be made between the formation of the one and of the other. If there is any principle in mineralogy, which may be considered as perfectly ascertained, it is, that rocks similarly stratified, and alternated with one another, are of the same formation.

Hence we conclude, that there is *no order of strata yet known*, that does not contain proofs of the existence of more ancient strata. We see nothing, in the strict sense, primitive. It must be understood, that what is here said has no reference to granite, which I do not consider as a stratified rock, and in which neither the remains of organized bodies, nor sand, have I believe been ever found; though some instances will be hereafter mentioned, where granite contains fragments of other stones, viz. of different kinds of primary schistus.

To the instances of sand involved in primary schistus, I might have added many from the rocks of that order on the coast of Berwickshire, of which mention is so often made in these Illustrations ; but I wished to draw the evidence from those rocks that are most unequivocally primary, and to which the Wernerian distinction of *intermediate* could not possibly be applied.

If any one assert, as M. Deluc has done, that sand is a chemical deposit, a certain mode of crystallization which quartz sometimes assumes, let him draw the line which separates sand from gravel ; and let him explain why quartz, in the form of sand, is not found in mineral veins, in granite, nor in basaltes, that is, in none of the situations where the appearances of crystallization are most general and best ascertained.

NOTE V. § 10.

Transportation of the Materials of the Strata.

156. The great transportation or *travelling* of the materials of the strata, supposed by Dr Hutton, has been treated as absurd by some of his opponents, particularly Deluc and Kirwan. These

philosophers seem not to have observed, that their own system, and indeed every system which derives the secondary strata from the primary, involves a transportation of materials, hardly less than is supposed in the Huttonian theory, and a degradation of the primeval mountains, in many instances much greater. To form some notion of this degradation, it must be recollected, that the primeval mountains, which furnished the materials of the secondary strata in the plains, cannot have stood in the place now occupied by these plains. This is obvious ; and therefore we must necessarily regard the secondary strata as derived from the primitive mountains which are the nearest to them, and of which a part still remains. This part is sufficient to define the base of the original mountains ; and the quantity of the secondary strata which surround them may help us to make some estimate of their height. Let us take, for instance, the extensive tract of secondary country about Newcastle, where coal mines have been sunk through a succession of secondary strata, to the depth of more than a thousand feet. This secondary country may be considered as comprehending almost the whole of the counties of Northumberland and Durham, and probably as extending very far under the part of the German Ocean which washes their coasts ; and the whole strata composing it must be derived, on the hypothesis we are now considering, from the

Cheviot Hills, on one side, and from those in the high parts of Westmoreland and Cumberland on the other, comprehending the Alston-Moor Hills, and the large group of primary mountains, so well known from the sublime and romantic scenery of the *Lakes*. Now, the mountains which stood on this base, had not only to supply the materials for the tract already mentioned, on the east, but had also their contingent to furnish to the plains on the west and north; the Cheviots to Roxburghshire and Berwickshire; the Northumberland mountains to the coal strata about Whitehaven, and along the sea coast to Lancashire. On the whole, we shall not exceed the truth, if we suppose, that the secondary strata, at the feet of the above mountains, are six or seven times more extensive than the base of the mountaincus tract. If then we take the medium depth of these secondary strata to be one thousand feet, it is evident, that the mass of stone which composes them, if it were placed on the same base with the primitive mountains, would reach to the height of six thousand feet. This is supposing the mass to preserve the breadth of its base uniformly to the summit; but if it be supposed to taper, as mountains usually do, we must multiply this six thousand by three, in order to have the height of these primeval mountains, which, therefore, were originally elevated not less than eighteen thousand feet: in height, therefore, they once rivalled the

Cordilleras, and are now but poorly represented by the hills of Skiddaw and Helvellyn. It were easy to show, that this estimate is still below the result that strictly follows from the Neptunian hypothesis ; but it is unnecessary to proceed further, than to prove, that the principle of the degradation of mountains, is involved in that hypothesis to an excessive and improbable degree ; and that the supporters of it, have either been guilty of the inconsistency of refusing to Dr Hutton the moderate use of a principle, which they themselves employ in its utmost extent, or of not having sufficiently adverted to the consequences of their own system.

157. The formation of secondary strata from the degradation of the contiguous mountains, on close examination, is subject to many other difficulties of the same kind. Mountains of secondary strata, and nearly horizontal, are found in this island of the height of three thousand feet. Such are Ingleborough, Wharnside, and perhaps some others on the west of Yorkshire. The whole chain, indeed, for secondary mountains, is of great elevation. The strata are of limestone, and of a very coarse-grained sandstone, alternating with it. No mountains can more clearly point out, that the strata of which they consist were once continued quite across the vallies which now separate them ; and hence, if the materials of those strata were indeed furnished from any contiguous primitive moun-

tains, the latter must have been, out of all proportion, higher than any mountains now in Britain.

158. Thus, a great degradation of the primitive
mountains, and of course a great travelling of their
materials, is proved to make a necessary part of the
Neptunian theory. The extent of this travelling
or transportation may be rendered more evident, if
we apply a similar mode of reasoning to larger portions of the globe. The north-west of Europe furnishes us an instance of a very extensive tract of
secondary country, comprehending the greater part
of Britain, the whole of Flanders and Holland,
part of Germany, the northern provinces of France,
and probably the bed of the German Ocean, at
least for a great extent. Within this circle almost
all is secondary, and on the sides of it all round
are placed ridges or groups of primitive mountains,
namely the mountains of Auvergne, at least in
part, and going round by the east, the Alps, the
Vosges, the Hartz, the Highlands and Western
Islands of Scotland, the hilly countries of Cumberland, Wales, and Cornwall. This zone of primitive mountains, on the supposition of the Neptunists, must have risen up in the form of islands in
the great ocean, that originally covered the earth,
forming a kind of circular Archipelago, including
in its bosom a sea, which was from seven to five
hundred miles in diameter. Over the whole of
this extent, the *detritus* of the above mountains

must have been carried, in order to form the flat interjacent countries which are now exposed to our view. Such then, even on their own supposition, is the extent to which the Neptunists must admit that the materials of the primeval mountains were transported by the ocean.

159. This transportation of materials, may not be so great as that which is involved in Dr Hutton's theory, but is such as should make the enemies of his system consider, how nearly the principles they *must* introduce, agree with those that they *would* reject. This is one fact, out of many, which shows, that there is at present a much nearer agreement between the systems of geology, than between their authors.

160. To these facts, demonstrating the great transportation of fossils in some former conditions of the globe, we may add another, recognised by all mineralogists. The animal *exuviæ* contained in limestone and marble, are often known to belong to seas, extremely remote from the countries where they are now found. In the chalk-beds of England, in the limestones of France, a great proportion of the petrifactions belong to the tropical seas, and appear to have been brought from the vicinity of the equator. Buffon observes, that of the fossil shells found in France, it has been disputed, whether the foreign are not more numerous than the native; and, though he is himself of opinion that

they are not, it is evident that they must bear a considerable proportion to the whole. * In the petrifactions of Monte Bolca, near Verona, where the impressions of fish are preserved between the laminæ of a calcareous schistus, one hundred and five different species have been enumerated, of which thirty-nine are from the Asiatic seas, three from the African, eighteen from those of South, and eleven from those of North America. † Similar observations have been made on the marine plants, and the impressions of vegetables, found in rocks, in different parts of Europe. At St Chaumont, near Lyons, is found an argillaceous schistus, covering a bed of coal, every lamina of which is marked with the impressions of the stem, leaf, or other part of some plant ; and it happens, says M. Fontenelle, by an unaccountable destination of nature, that not one of these plants is a native of France. They are all ferns of different species, peculiar to the East Indies, or the warmer climates of America. Here also was found the fruit of a tree, which grows only on the coasts of Malabar and Coromandel. ‡

The same holds of the bodies of amphibious ani-

---

* Buffon, Théorie de la Terre, Art. 8.

† Saussure, Voyages aux Alpes, Tom. III. § 1535.

‡ Mém. de l'Acad. des Sciences, 1718, p. 3. and 287 ; and 1721, p. 89, &c.

mals which now make a part of the fossil kingdom. The head and the bones of crocodiles have been found in the island of Sheppey, at the mouth of the Thames ; and the remains of an animal of the same species, but of a variety now peculiar to the Ganges, have been discovered in the alum rocks on the coast of Yorkshire. * These proofs of the transportation of materials by the sea, have the advantage of involving nothing hypothetical, and of being equally addressed to the geologists of every persuasion.

On this subject I cannot help observing, that the accurate comparison of the animal exuviæ of the mineral kingdom, with their living archetypes, is not merely a curious inquiry, but is one that may lead to important consequences, concerning the nature and direction of the forces which have chang-

---

* Phil. Trans. Vol. I. p. 688. Camper denies that the remains here mentioned belong to the crocodile, or any amphibious animal, and refers them to the balæna. He passes the same judgment on those fossil bones from St Peter's Mount, near Maestricht, which have been supposed to belong to the crocodile ; he looks on them as belonging to whales, though of an unknown species. In this Mount, so famous for its petrifactions, he finds many specimens of bones, which he thinks belong to the turtle. Phil. Trans. Vol. LXXVI. p. 443. The opinion of an author, so well skilled in comparative anatomy, must be regarded as of great weight: if it takes from our argument in one part, it adds to it in another, and the acquisition of the turtle makes up abundantly for the loss of the crocodile.

ed, and are continually changing, the surface of the earth.

161. These remarks I have thought it proper to add to the proofs of the composition of the present from former strata, in order to show, that the great transportation of materials involved in that supposition, is not only conformable to the hypothesis of the Neptunists concerning the secondary strata, but is also proved by the most direct evidence, independently of all hypothesis. All this reasoning regards the ancient state of the globe. Whether such a travelling of stony bodies makes a part of the system now actually carrying on, will be considered in another place. *

NOTE VI. § 13.

Mr Kirwan's Notion of Precipitation.

162. The Neptunist who has provided the means of dissolving the materials of the strata, has only performed half his work, and must find it a task of equal difficulty to force this powerful menstruum to part with its solution. Mr Kirwan, aware in some degree of this difficulty, has attempted to obviate it in a very singular way. First, he ascribes the so-

---

* See NOTE XIX.

lution of all substances in water, or in what he calls the chaotic fluid, to their being finely pulverised, or created in a state of the most minute division. Next, as to the deposition, the solvent being, as he acknowledges, very insufficient in quantity, the precipitation took place, (he says,) on that account the more rapidly.

If he means by this to say, that a precipitation without solution would take place the sooner the more inadequate the menstruum was to dissolve the whole, the proposition may be true ; but will be of no use to explain the crystallization of minerals, (the very object he has in view,) because to crystallization, it is not a bare subsidence of particles suspended in a fluid, but it is a passage from chemical solution to non-solution, or insolubility, that is required.

If, on the other hand, he means to say, that the solution actually took place more quickly, and was more immediately followed by precipitation, because the quantity of the menstruum was insufficient, this is to assert, that the weaker the cause, the more instantaneous will be its effect.

Of two propositions the one of which is nugatory, and the other absurd, it is not material to inquire which the author had in view.

NOTE VII. § 16.

Compression in the Mineral Regions.

163. It is worthy of remark, that the effects
ascribed to compression in the Huttonian Theory,
very much resemble those which Sir Isaac Newton supposes to be produced in the sun and the
fixed stars by that same cause. " Are not," says
he, " the sun and fixed stars great earths, vehemently hot, whose heat is conserved by the greatness of the bodies, and the mutual action and reaction between them, and the light which they
emit; *and whose parts are kept from fuming
away, not only by their fixity, but also by the
vast weight and density of the atmospheres incumbent upon them, and very strongly compressing
them.*" *

164. The fact of water boiling at a lower temperature under a less compression, is sufficient to
justify the supposition, that bodies may be made by
pressure to endure extreme heat, without the dissipation of their parts, that is, without evaporation
or combustion. A further *postulatum* is introduced in Dr Hutton's theory, namely, that compound bodies, such as carbonate of lime, when the

---

* Newton's Optics, Query 11.

compression prevents their separation, may admit of fusion, notwithstanding that the fixed part may be infusible when separated from the volatile. This assumption is supported by the analogical fact of the fusion of the carbonate of barytes, as mentioned in the text.

165. In a region where the action of heat was accompanied with such compression as is here supposed, there could be no fire, properly so called, and no combustion; this is admitted by Dr Hutton, and it is therefore a fallacious argument which is brought against his theory, from the impossibility of fire being maintained in the bowels of the earth. This impossibility is precisely what he supposes; and yet Mr Kirwan's arguments are directed, not against the existence of heat in the interior of the earth, but against the existence of burning and inflammation.

After taking notice,* that Saussure had succeeded, though with extreme difficulty, in melting a particle of limestone, so small as to be visible only with a microscope, " what (adds he) must have been the heat necessary to melt whole mountains of this matter? Judging by all that we at present know of heat, such a high degree could only be produced by the purest air, acting on an

---

* Geol. Essays, p. 453.

8

enormous quantity of combustible matter. Now, Ehrman observed, that the combustion of two hundred and eighty cubic inches of air, acting on charcoal, was not able to effect the fusion of one grain of Carrara marble ; from whence it is apparent, that all the air in the atmosphere, nor in ten atmospheres, would not melt a single mountain of this substance, of any extent, even if there were a sufficient quantity of inflammable matter for it to act upon. Judging also of subterraneous heat by what we know of that of volcanoes, no such heat exists : the highest they in general produce, is that requisite for the fusion of the volcanic glass called obsidian, which Saussure found not to exceed 115° of Wedgewood ; but basaltine, which requires 140° of Wedgewood, is never melted in the lavas of Ætna. How little capable, then, would volcanic heat be to effect the fusion of Carrara marble, which, according to the same excellent author, would require a heat of upwards of 6300° of Wedgewood, if this pyrometer could extend so far ? And in what circumstances does Dr Hutton suppose this astonishing heat to have existed, and even still to exist, under the ocean, in the bowels of the earth, where neither a sufficient quantity of pure air, nor of combustible matter, capable of such mighty effects, can, with any appearance of probability, be supposed to exist ; and,

12

without these, such degrees of heat cannot even be imagined, without flying into the region of chimeras."

166. Now, this reasoning is not applicable to Dr Hutton's hypothesis of subterraneous heat, because it is grounded on experiments, where that very separation of the volatile and fixed parts takes place, which is excluded in that hypothesis. When limestone or marble is exposed to such heat as is here mentioned, or even to heat of a degree vastly inferior, the carbonic gas is expelled, and the body is reduced to pure lime; from the refractory nature of which, as we learn from the fact relative to barytes, mentioned above, no conclusion can be drawn as to the infusibility of the same substance, when combined with the carbonic gas. The Carrara marble may require a heat of 6300° of Wedgewood, to melt it in the open air, where the carbonic gas escapes from it; but under such a pressure as would retain this gas, it cannot be inferred, that it might not melt with the heat of a glass-house furnace. In like manner, it may be true, that two hundred and eighty cubic inches of air, acting on charcoal, cannot effect the fusion of one grain of this marble, after its fixed air is driven off from it; but we cannot from thence draw any inference, applicable to a case where the carbonic gas is retained, and where the action of heat is independent of atmospheric air.

Nothing, therefore, can be more inconclusive than this reasoning, as it proceeds on the supposition, that Dr Hutton's system admits propositions, which in fact it expressly denies.

167. Of the production and maintenance of heat, in circumstances so different from those of ordinary experience, we can hardly be expected to give any explanation; but we are not entitled, merely on that account, to doubt of the existence of such heat. Mr Kirwan thinks otherwise: " Judging," he says, " from all we at present know of heat, such a high degree of it, (as will melt limestone,) could only be produced by the purest air, acting on an enormous quantity of combustible matter. Without these, such degrees of heat cannot even be imagined, without flying into the region of chimeras." *

Now, in the first place, the high degree of temperature which is here understood, is probably not necessary to the purposes of mineralization, as has just been shown; and, in the second place, it is not FIRE, in the usual sense of the word, but HEAT, which is required for that purpose; and there is nothing *chimerical* in supposing, that nature has the means of producing heat, even in a very great degree, without the assistance of fuel or of vital air. Friction is a source of heat, unlimited, for what we

---

* Geol. Essays, p. 454.

know, in its extent, and so perhaps are other operations, both chemical and mechanical; nor are
either combustible substances, or vital air, concerned in the heat thus produced. So also the heat of
the sun's rays in the focus of a burning glass, the
most intense that is known, is independent of the
substances just mentioned; and, though that heat
certainly could not calcine a metal, nor even burn
a piece of wood, without oxygenous gas, it would
doubtless produce as high a temperature in the absence as in the presence of that gas.

It is true, that it is not by the solar rays that
subterraneous heat is produced; but still, from
this instance, we see, that there is no incongruity
in supposing the production of heat to be independent of combustible bodies, and of vital air. We
are indeed, in all cases, strangers to the origin of
heat; philosophers dispute, at this moment, concerning the source of that which is produced by
burning; and much more are they at a loss to determine, what upholds the light and heat of the
great luminary, which animates all nature by its
influence. If we would form any opinion on this
subject, we shall do well to attend to the suggestions of that great philosopher, who was hardly less
distinguished from others by his doubts and conjectures, than by his most rigorous and profound
investigations. " May not great, dense, and fixed
bodies, when heated beyond a certain degree, emit

light so copiously, as, by the emission and reaction of its light, and the reflections and refractions of its rays within its pores, to grow still hotter, till it comes to a certain period of heat, such as is that of the sun ?   And, are not the sun and fixed stars great earths, vehemently hot, whose heat is conserved by the greatness of the bodies, and the mutual action and reaction between them and the light which they emit ?" *

168. Some recent experiments, seem to make the suggestions in this query applicable to an opaque body like the earth, as well as to luminous bodies, such as the sun and fixed stars.   The radiation of heat, where there is no light, was first rendered probable by the experiments of M. Pictet of Geneva ; † and the only objections to which the conclusions from those experiments seemed liable, are removed by the late very important discoveries of Dr Herschel. ‡   From these it appears, that heat is capable of refraction and reflection, as well as light, so that it is not absurd to suppose, that *the heat of great, dense, and fixed bodies, may be conserved by the greatness of the bodies, and the mutual action and reaction between them and the heat which they emit.*

---

* Newton's Optics, *ubi suprà.*
† Essai sur le Feu.
‡ Phil. Trans. 1800, p. 84.

The existence of subterraneous heat is still further rendered probable from the researches of
Mairan, which tend to show, that there is another
source of terrestrial heat besides the influence of
the solar rays. *

Whatever be the truth with regard to these conjectures, it is certain, that the first and original
source of heat is independent of burning. Burning is an *effect* of the concentration of heat; and
though, by a certain reaction, it has the power of
continuing and augmenting that heat, it never can
be regarded as its primary and material cause.
When, therefore, we suppose a source of heat, independent of fire and of burning, we suppose what
certainly exists in nature, though we are not informed of the manner of its existence, nor of its
place, otherwise than from considering the phenomena of the mineral kingdom.

169. Lastly, we are not entitled, according to
any rules of philosophical investigation, to reject a
principle, to which we are fairly led by an induction from facts, merely because we cannot give a
satisfactory explanation of it. It would be a very
unsound view of physical science, which would induce one to deny the principle of gravitation,
though he cannot explain it, or even though the
admission of it reduces him to great metaphysical

---

* Mém. de l'Acad. des Sciences, 1765, p. 143.

difficulties. If indeed a downright absurdity, or inconsistency with known and established facts, be involved in any principle, it ought not to be admitted, however it may seem calculated to explain other appearances. If, for instance, Dr Hutton held, that combustion was carried on in a region where there was no vital air, we should have said, that he admitted an absurdity, and that a theory founded on such *postulata* was worse than chimerical. But, if the only thing imputable to him is, that, being led by induction to admit the fusion of mineral substances in the bowels of the earth, he has assumed the existence of such heat as was sufficient for this fusion, though he is unable to assign the cause of it, I believe it will be found, that his system only shares in an imperfection, which is common to all physical theories, and which the utmost improvement of science will never completely remove.

170. Thus, then, we are led, it must be allowed, into the *region of hypothesis and conjecture*, but by no means into that of *chimeras*. Indeed, the reproach of flying into the latter region, may be said to come but ill from one, who has trode so often the *crude consistence* of the chaos, and who delights to dwell beyond the boundaries of nature. By sojourning there long, it is not impossible that the eye may become so accustomed to fantastic

forms, that the figures and proportions of nature shall appear to it deformed and monstrous.

NOTE VIII. § 24.

Sparry Structure of Calcareous Petrifactions.

171. When the shells and corals in limestone are quoted by mineralogists, it is not always considered in what state they are found. In general, they have a sparry structure, very different from that of the original shell or coral, of which, however, they retain the figure with wonderful exactness, though probably sometimes altered in size. Though sparry, they are often foliated, and preserve their animal, in conjunction with their mineral, texture. Now, this crystallization is a mark of some operation, quite different from any that can be ascribed to the water in which these bodies had their origin, and by which they were brought into their place. They were impervious to water; and it cannot be said that their sparry structure has been derived from the percolation of that fluid, carrying new calcareous matter into their pores. We can account for the change produced in them, I think, only by supposing them to have been softened by heat, so as to permit their parts to arrange

selves anew, and to assume the characteristic organization of mineral substances.

All shells have not the change effected on them
that is here referred to ; those in chalk, for instance, retain very much their original form in all
respects.   This is what we might expect from the
very different degree of intensity, with which the
mineralizing cause has acted on chalk, and on limestone or marble.   In general, it is in the hardest
and most consolidated limestone, that the marine
objects are most completely changed into spar.

It would be exceedingly interesting to examine,
whether any of the phosphoric acid remains united
to shells of either of these kinds.   We might most
readily expect it to be united, in a certain degree,
to the shells that are least mineralized.

This experiment would enable us also to appreciate the force of Mr Kirwan's argument against
the finer marbles, such as the Carrara, containing
shells. *   This argument proceeds on an experiment, mentioned in the Turin Memoirs for 1789,
from which it appears, that no phosphoric acid is
found in pure limestone ; and its absence, Mr Kirwan says, cannot be attributed to fusion, as phosphoric acid is indestructible by heat.

He calls this a demonstration ; but, in order to

---

* Geol. Essays, p. 458.

entitle it to that name, it will be necessary, first, to prove, that phosphoric acid exists in those lime-stones which evidently consist of shells in a mine-ralized state. If these are found without phos-phoric acid, it is evident that the preceding argu-ment fails entirely. If they are found to contain that acid, it will then no doubt afford a probability, though not a demonstration, that Carrara marble does not directly originate from shells.

That nature has some process, by which the above acid is separated from the earth of bones, and probably also from the earth of shells, is evi-dent from the state in which the bones are found in the caves of Bayreuth. Those that are the most recent, and least petrified, contain most of the phosphoric acid. Where the petrifaction has proceeded far, that acid is not found.

172. Among many of the strata, such a fluidity has prevailed, as to enable some of the substances included in them to crystallize. Calcareous spar and siliceous crystals are often found in stratified rocks, forming veins of secretion, or lining close cavities, included on all sides by the uncrystallized rock. In the instances of gneiss, and many spe-cies of marble, almost the whole matter of the stratum is crystallized. This union of a stratified and crystallized structure in the same substance, has a great affinity to that union of the crystallized with the *organic* structure of shells and corals

which has just been mentioned; and both are
doubtless to be referred to the same cause.

NOTE IX. § 31.

Petroleum, &c.

173. According to the theory of coal laid down
above, its two chief materials, charcoal and bitumen, being furnished by the vegetable and animal
kingdoms, both of the land and of the sea, have
formed with one another a new combination, by
the action of subterraneous heat; but have also, in
some cases, been separated by that same action,
where the degree of compression necessary for
their union, happened to be wanting. The carbonic part, when thus separated from the bituminous, forms an infusible coal, which burns without
flame: the bituminous part, when separated from
the carbonic, is found in the various states of
naphtha, petroleum, asphaltes, and jet.

The great resemblance of infusible or blind
coal, to the residuum obtained by the distillation
of bituminous coal; and again, the coincidence of
the bitumens just named, with the volatile part, or
the matter brought over by such distillation, are
strong arguments in favour of this theory. The
other facts in the natural history of coal, serve to

confirm the same conclusion ; but it must be confessed, that what we know of the pure bitumens,
except the circumstance just mentioned, is of a
more ambiguous nature, and may be reconciled
with different theories. The drops of petroleum
contained within the cavities of the limestone,
mentioned at § 31, are however strong facts in
confirmation of Dr Hutton's opinions, and they
are furnished by the substances purely bituminous. A careful examination would probably
make us acquainted with others of the same kind,
for limestone is very often the matrix in which
petroleum and asphaltes are contained. The
greatest mine of asphaltes in Europe, that in the
Val de Travers, in the territory of Neufchâtel, is
in limestone, from which, though it in some places
exudes, it is in general extracted by the application of heat. The strata for several leagues are
impregnated with bitumen ; and, if examined with
attention, would probably afford specimens similar
to those which have just been mentioned.

174. It is a general remark, that, where petroleum is found, on digging deeper, they come to
asphaltes ; and, at a depth still greater, they discover coal. This probably does not hold invariably ; but it is certain, that most of the fountains of petroleum are in the neighbourhood of
coal strata. Petroleum and asphaltes are found in
great abundance in Alsace, in a bed of sand, be-

tween two beds of clay or argillaceous schistus, and the same country also affords coal. *   This is true likewise of the fossil pitch of Coalbrook-dale ; and of the petroleum found in St Catharine's Well, near Edinburgh.   Auvergne † contains abundance of fossil pitch, which exudes, in the warm season, from a rock impregnated with it through its whole mass.   There are also coal strata in the same country, not far distant.

A very satisfactory observation relating to this subject, has lately been communicated from a country, with whose natural history we were till of late entirely unacquainted.   In the Burmha empire, petroleum is dug up in an argillaceous earth, from the depth of seventy cubits.   This argilla‐ ceous earth, or schistus, lies under a bed of free‐ stone ; and under all, about one hundred and thirty cubits from the surface, is a bed of coal. ‡

175. In the petroleum lake of the Island of Trinidad, described *Phil. Trans.* 1789, the petro‐ leum evidently exudes from the rock, and is col‐ lected in a variety of springs in the bottom, after which it hardens, and acquires the consistency of pitch.   The manner, therefore, in which petro‐ leum exists in the strata, is very consistent with

----

* Encyclopédie, mot, *Asphalte.*
† Voyage en Auvergne, par Legrand, Tom. I. p. 351.
‡ Asiatic Researches, Vol. VI. Art. 6. p. 130.

the idea of its having been introduced in the form
of a hot vapour.

Even amber appears to have some relation to
coal. It is found in the unconsolidated earth in
Prussia and Pomerania; but I am not sure whether
this earth is *travelled* or not. In the same earth
where the amber is found, there is often a mixture
of coaly matter, which burns in the fire; it is apparently fibrous, and has been considered as a kind
of fossil wood. *

These circumstances make out a connection between the purer bitumens and ordinary coal; but
do not, it must be acknowledged, establish any
thing with respect to the more immediate relation,
supposed in this theory to exist between them and
blind coal. It is probable, indeed, that, to discover any facts of that kind, the natural history of
both substances must be more carefully examined;
the natural history of blind coal, in particular, has
hitherto been but little attended to.

176. A fact is mentioned by Mr Kirwan, which
must not be regarded as less valuable for being adverse to this theory. It is, that neither petroleum,
nor any fossil bitumen, is found in the vicinity of
the Kilkenny coal, as might be expected, if that
coal was deprived of its bituminous part by subter-

---

* Buffon, Hist. Nat. des Mineraux, Tom. II. p. 5.

raneous distillation. * This, however, admits of explanation. Though a general connection, on the above hypothesis, might be expected between bitumens and infusible coal, we cannot look for it in every instance. The heat which drove off the bitumen from one part of a stratum of coal, may only have forced it to a colder part of the same stratum ; and thus, in separating it from one portion of carbonic matter, may have united it to another. Blind coal may therefore be found where no bitumen has been actually extricated. In like manner, bitumen may have been separated, where the coal was not reduced to the state of coak, as a part of the bitumen only may have been driven off, and enough left to prevent the coal from becoming absolutely infusible.

It should be considered too, if the bitumen was really separated, and forced, in the state of vapour, into some argillaceous or limestone stratum, that this stratum may have been wasted and worn away long ago, so that the bitumen it contained may have entirely disappeared. It does not therefore necessarily follow, that, wherever we find blind coal, there also we should discover some of the purer bitumens.

---

* Geol. Essays, p. 473.

NOTE x. § 37.

*The Height above the Level of the Sea at which
the Marks of Aqueous Deposition are now
found.*

177. We have two methods of determining the
minimum of the change which has happened to the
relative level of the sea and land ; or for fixing a
limit, which the true quantity of that change must
necessarily exceed. The one is, by observing to
what height the regular stratification of mountains
reaches above the present level of the sea; the
other is, by determining the greatest height above
that level, at which the remains of marine animals
are now found. Of these two criteria, the first
seems preferable, as the fact on which it proceeds
is most general, and least subject to be affected by
accidental causes, or such as have operated since
the formation of the rocks. The results of both,
however, if we are careful to select the extreme
cases, agree more nearly than could have been expected.

178. The mountain Rosa, in the Alps, is entirely of stratified rocks, very regularly disposed,
and nearly horizontal. *   The highest summit of

---

* Voyages aux Alpes, Tom. IV. § 2138.

this mountain is, by Saussure's measurement, 2430 toises, or 14739 English feet, above the level of the sea, or lower than the top of Mont Blanc only by 20 toises, or 128 feet. * This is, I believe, the highest point on the earth's surface, at which the marks of regular stratification are certainly known to exist ; for though, by the account of the same excellent mineralogist, Mont Blanc itself is stratified, yet, as the rock is granite, the stratification vertical, and somewhat ambiguous, it is much less proper than Monte Rosa for ascertaining the limit in question.

179. Again, in the new Continent, we have an instance of shells contained in a rock, not much lower than the summit of Monte Rosa. This is one described by Don Ulloa, near the quicksilver mine of Guanca-Velica, in Peru. The height at which a specimen of these shells, given by Ulloa to M. Legentil, was found, was $2222\frac{1}{3}$ toises, or 14190 feet English, above the level of the sea. † This height agrees with the preceding, within 549 feet, a quantity comparatively small.

180. The last of the facts just mentioned is curiously commented on by Mr Kirwan. As he has proved, he says, that the mountains higher than

---

* Voyages aux Alpes, Tom. IV. § 2135.

† See Hist. Acad. des Sciences, 1770. Phys. Générale, No. 7.

8500 feet were all formed before the creation of fish, it follows, that the shells found at Guanca-Velica, must have been carried there by the deluge.* Now, without objecting to the proof here referred to, (though it seems very open to objection,) it is sufficient to remark, that, if the shells at Guanca-Velica were carried there by the deluge, or any other cause that operated after the formation of the rock of which the mountain consists, they can make no part of that rock, but must lie, like other adventitious fossils, loose and detached on the surface, or at most externally agglutinated to the stone. This, however, is certainly not the fact ; for, in the account just quoted, we read, that Don Ulloa told M. Legentil, " qu'il avoit détaché ces coquilles d'un banc fort épais." This seems plainly to indicate, that the shells were included in a bed of rock. But, granting that the expression is a little ambiguous, on turning to the *Mémoires Philosophiques* of the same author, the difficulty is completely removed, and it is made evident, that these shells are in fact integrant parts of the rock. " On voit dans ces montagnes-là, (about Guanca-Velica, and particularly at that in which is the quicksilver mine,) des coquilles entieres, petrifiées et enfermées au milieu de la roche, que les eaux de pluie mettent à decouvert. Ces coquilles font corps avec la pierre ;

---

* Geol. Essays, p. 54.

mais malgré cela, on remarque que la partie qui fut
coquille, se distingue par la couleur, la structure,
la qualité de la matière de tout autre corps pierreux
qui l'enferme, et du massif qui s'est fixé entre les
deux ecailles," * &c.   He goes on to say, that one
can distinguish marks of these shells having been
worn, before they were included in the stone.

181. Thus it appears, that whatever proof any
fossil shell affords, that the rock in which it is found
was formed under the sea, the same is afforded by
the fossil shells of Guanca-Velica; and we are,
therefore, perfectly entitled to conclude, that the relative level of the sea and land has changed, since
the formation of the latter, by more than 14000
feet.   The height assumed in § 37 is therefore
much under the truth; and the water, for which
the Neptunists must provide room in subterraneous
caverns, might very well have been stated at more
than a five-hundredth part of the whole mass of the
earth.

Thus also the argument by which the Neptunists would connect the creation of fish with the beginning of the secondary mountains, falls entirely
to the ground.   Indeed, it is strange that Mr Kirwan should have supposed it possible, that the shells
in question were loose and unconnected with the

---

* Mém. Philosophiques de Don Ulloa, Discours XVI.
Vol. I. p. 364.

rock, and had continued so, ever since the deluge, in such elevated ground, where the torrents wear and cut down the mountains with unexampled violence, and have hollowed out *Quebradas* so much deeper and more abrupt than the glens or vallies among other mountains.   He had not, I believe, seen the passage I have quoted from Ulloa ; but the circumstances did not warrant the shells in question to be regarded as extraneous and adventitious fossils.   A geologist should have known better than to suppose this possible.   When we see Voltaire ascribing to accidental causes the transportation of those shells which he had been told were often found among the Alps, we can excuse in a Poet and a Wit, that ignorance of the facts in mineralogy, which concealed from him the extreme absurdity of his assertion ; but when a Chemist or Mineralogist talks and reasons in the same manner, we cannot consider him as entitled to the same indulgence.

NOTE XI. § 42.

Fracture and Dislocation of the Strata.

182. The greatest part of the facts relative to the fracture and dislocation of the strata, belongs to the history of veins.   The instances of *slips,* where no

new mineral substance is introduced between the
separated rocks, are what properly belong to this
place.   The frequency of these, and their great extent, are well known wherever mines have been
wrought.   In some of them no opening is left, but
the slipped strata remain contiguous; in other
cases, there is introduced an unconsolidated earth,
often a clay, which may be supposed to have come
from above, and very probably to have been carried
down by the water.   In some such cases, however,
there are not wanting appearances, which show the
matter in the slip to have been forced up from below, as we find it to contain substances which could
not have come from the surface. *

183. A very remarkable fact of this kind occurred not long ago, in digging the Huddersfield
canal in Yorkshire; and a very distinct account of
it is given in the *Philosophical Transactions*, by
the engineer who directed the work.   In carrying
a tunnel into the heart of a hill, the miners came
to what is called in the description a *fault*, *throw*,
or *break*, or what we have here called a shift,
which was filled with *shale* set on edge, mixed
with softer earth, and in some places with small
lumps of coal.   The fault or space filled with

---

* Unconsolidated earth contained between the sides of a
rock that has slipped, is frequent in Cornwall, and is called a
Fleukan.

these materials, was in general about four yards
broad, and lay nearly in the direction of the tunnel, so that a considerable extent of it was visible·
Beside the shale, it contained a *rib* of limestone,
about four feet thick, which run parallel to the
sides of the *fault*, and about four feet from the
southern margin of it. On each side of this rib
were found balls of limestone, promiscuously scattered, and of various sizes, from an ounce to one
hundred pounds weight. The balls, when broken,
were found to contain some pyrites near their
edges; they were not perfectly globular, but flat-
tened on the opposite sides, and similar to one
another. * At the time when the account was
written, about seventy yards of the *rib* had been
discovered.

184. Now, it is certain, that neither this rib of
limestone, nor the balls that accompanied it, can
have come from above, as there is no limestone
within twenty miles of the place where they were
found. They must, therefore, have been forced
up from below, and no doubt belong to some limestone strata, which lie there at a great depth under the surface. The length of this fragment of
rock, which, from the account, one must suppose
to have been entire, conveys no mean idea, either
of the intensity or regularity of the force by which

---

* Phil. Trans. 1796, p. 350.

it was brought into its present situation.   In veins,
it is not uncommon to meet with stones that appear to have come from a greater depth : but this
is probably the most remarkable instance of the
same phenomenon, which has appeared in a mere
slip, and none, I think, can speak a language less
liable to be misunderstood.

185. I shall here mention another mark of
violent fracture, that has been observed in rocks of
breccia or pudding-stone, which, though not of
the same kind with the preceding, and of a nature
quite peculiar, belongs rather to this place than
any other.   In rocks of the kind, just mentioned,
it sometimes happens, that considerable portions
are separated from one another, as if by a mathematical plane, which had cut right across all the
quartzy pebbles in its way.   None of the pebbles
are drawn out of their sockets, that is, out of the cement that surrounds them, but are divided in two
with a very smooth and even fracture. The pebbles,
in the instances which I have seen, were of quartz,
and other species of primary and much indurated
rock.

Lord Webb Seymour and I observed pudding-
stone rocks, exhibiting instances of this singular
kind of fracture, near Oban, in Argyleshire, about
three years ago.   The phenomenon was then entirely new to us both ; but I have since met with
an instance of the same kind in Saussure's last

work. As the fact is of so particular a kind, I shall state it in his own words : The place was on the sea shore, near the little town of Alassio, between Nice and Genoa.

" En passant entre ces blocs de brèche, j'admirai quelques-uns d'entr'eux, d'une grandeur considerable, et taillés en cubes, avec la plus parfaite régularité. Il y avoit ceci de remarquable, c'est que l'action de la pesanteur, qui avoit taillé ces cubes en rompant leurs couches, avoit coupé tous les cailloux des brèches à fleur de la surface de la pierre, aussi nettement que si c'eût été une masse molle qu'on eût tranchée verticalement avec un rasoir. Cependant parmi ces cailloux, la plupart calcaires, il s'en trouvoit de très durs, de petrosilex, par exemple, même de jade, qui étoient tranchées tout aussi nettement que les autres." *

186. This description is no doubt accurate, though it involves in it something of theory, viz. that the fracture was made by the weight of the stone. This may indeed be true : the operation probably belongs altogether to the surface, and is one with which the powers of the mineral regions are not directly concerned. The phenomenon, however, appears to me, on every supposition, very difficult to explain. In the specimen which I brought from Oban, the smallest pieces of stone

---

* Voyages aux Alpes, Tom. iii. § 1731.

are cut in two, as well as the largest. The consolidation and hardness of the mass are very great, and the connection of the different frag- ments so perfect, that it is no wonder the whole should break as one stone. But still, that the fracture should be so exactly in one plane, and without any shattering, is not a little enigmatical ; if it is indeed a fracture, it must be the consequence of an immense impulse, very suddenly communi- cated.

NOTE XII. § 43.

Elevation and Inflection of the Strata.

187. The evidence of the different formation of the primary and secondary strata, and of the changes which the former have undergone, is best seen at the points where those strata come into contact with one another. Dr Hutton was not the first who observed these junctions, though the first who rightly interpreted the appearances which they ex- hibit. He has mentioned observations of this sort by Deluc on the confines of the Hartz ; by the au- thor of the *Tableau de la Suisse*, at the pass of Yetz ; by Voight, in Thuringia ; and Schreiber, at the mountain of Gardette. *

---

* Theory of the Earth, Vol. I. p. 410, to 453.

The leading facts to be remarked, are,

I. The vertical or very upright position of the primary or lower strata.

II. The superstratification of the secondary, in a position nearly horizontal, so as to be at right angles to those on which they rest.

III. The interposition of a breccia between them ; or, as happens in many cases, the transition of the lowest of the secondary beds into a breccia, containing fragments sometimes worn, sometimes angular, of the primary rock.

This last is a phenomenon extremely general, and all our subsequent information confirms Dr Hutton's anticipations concerning it. " It will be very remarkable," he says, " if similar appearances, (such as those of the breccia described by Voight,) are always found upon the junction of the Alpine with the level countries." * Saussure, in a part of his work, not published when Dr Hutton wrote this passage, has attested the generality of the fact with respect to the whole Alps, from the Tyrol to the Mediterranean : " Un fait que l'on observe sans aucune exception, ce sont les amas de débris, sous la forme de blocs, de brêches, de poudingues, de grès, de sable, ou amoncelés, et formant des montagnes, ou des collines, dispersés sur le bord

---

* Theory of the Earth, Vol. I. p. 448.

exterieur, ou même dans les plaines qui bordent la chaine des Alpes." *

This passage is perfectly decisive as to the generality of the fact, that the Alps, from the Tyrol to the Mediterranean, are bordered all round by pudding-stones or breccias. At the same time, it is necessary to remark, that M. Saussure, by enumerating loose blocks and sand, along with pudding-stones, breccias and grit, confounds together things which are extremely different, and which have had their origin at periods extremely remote from one another. The consolidated rocks of breccia, pudding-stone and grit, though they are indications of waste, have received their present character at the bottom of the sea : the loose blocks of stone, the sand and gravel, on the other hand, are the effects of the waste now going forward on the surface of the land, and are the materials out of which rocks of the three kinds just mentioned may hereafter be composed. If so skilful a mineralogist as Saussure is guilty of such inaccuracy, it must be ascribed to the confusion necessarily arising from the system which he followed, and not to his own want of discrimination.

188. The same phenomenon, of a breccia circumscribing the primary mountains, is met with in

---

* Voyages aux Alpes, Tom. IV. § 2330.

Scotland ; and the Grampians, wherever they are bounded by secondary strata, whether on the south or north, afford examples of it. The breccia generally consists of the fragments of the primary rock, most commonly rounded, but sometimes also angular, united by a cement of secondary formation, and the whole disposed in horizontal beds. It was on the constancy of this accompaniment of the primary strata, and on the great quantity of highly polished gravel often included in these breccias, that Dr Hutton grounded the hypothesis of the double raising up and letting down of the ancient strata. See § 43.

189. As the spots where the primary and secondary rocks may be seen in contact with one another are of great importance in geology, and present to the senses the most striking monuments of the high antiquity and great revolutions of the globe, it may be useful to point out such of them as have been observed in this island. To those which Dr Hutton has described, I have a few more to add, the result of some geological excursions, which I made in company with the Right Honourable Lord Webb Seymour, to whose assistance I have been much indebted in the prosecution of these inquiries.

190. The most southern junction which we observed is at Torbay, where the ancient schistus which prevails along the coast, from the Land's End to that point, receives a covering of red hori-

zontal sandstone, the same which composes the greater part of Devonshire. The spot where the immediate contact is visible, is on the shore, a little to the south of Paynton ; and one circumstance, which among many others serves to distinguish the different formation of the two kinds of rock, is, that the schistus, which is elevated here at an angle of about 45°, is full of quartz veins, which veins are entirely confined to it, and do not, in as far as we could observe, penetrate into the sandstone, in a single instance. It is probable, that on the north shore of the bay, the same line of junction is visible : we saw it at Babicomb Bay, still more to the northward.

191. From this place, the secondary strata of different kinds prevail without interruption, along the coast of the British Channel, and of the German Ocean, as far as Berwick upon Tweed, and for some miles beyond it. The sea coast then intersects a primary ridge, the Lammermuir Hills, which traverses Scotland from east to west, uniting, near the centre of the country, with the metalliferous range of Leadhills, and afterwards with the mountains of Galloway. The section which the sea coast makes of the eastern extremity of this ridge, is highly instructive, from the great disturbance of the primary strata, and the variety of their inflections. The junction of these strata with the secondary, on the south side, is near the little

sea-port of Eyemouth, but the immediate contact is not visible.

On the north side of the ridge, the junction is at a point called the *Siccar*, not far from Dunglass, the seat of Sir James Hall, Baronet.   By being well laid open, and dissected by the working of the sea, the rock here displays the relation between the two orders of strata to great advantage.   Dr Hutton himself has described this junction ; Theory of the Earth, Vol. I. p. 464.

192. From the point just mentioned, the secondary strata continue as far as Stonehaven, where the southern chain of the Grampian mountains is intersected by the sea coast.   Here a great mass of pudding-stone appears to lie on the primary strata, but their immediate contact has not been observed.

193. Going along the coast toward the north, the next junctions which we saw were on the shore, one near Gardenston, and another near Cullen, in Banffshire.   The latter is very distinct ; it is about a mile to the westward of the rocks called *The Three Kings*, where a red sandstone, the lower beds of which involve much quartzy gravel, lies horizontally upon very regular, upright, and highly indurated strata.   Some of these strata are micaceous, and others of the granulated quartz, mentioned in § 153.

194. This last is, I believe, the most northern junction which has been observed in our island.

The western coast furnishes several more, which however are not all visible. The line of separation, between the primary schistus of the Grampians and the sandstone which covers it, is intersected at its western extremity by the Frith of Clyde, not far from Ardencaple in Dunbartonshire. The two kinds of stone can be traced within a few yards of each other, but not to the actual contact : the beds of sandstone nearest the schistus form as usual a breccia, loaded with fragments of the primary rock. The secondary rock, which begins here, continues for about fifty miles south, to Girvan in Ayrshire, where the primary schistus again rises up, but is not seen in contact with the secondary. It extends to the Mull of Galloway and the shores of the Solway Frith.

The Isle of Arran, however, not far distant from this part of the coast, contains a junction at its northern extremity, where secondary strata of limestone lie immediately on a primary micaceous schistus. This is described by Dr Hutton, and was the first phenomenon of the kind which he had an opportunity of examining. * The junction is visible but at one spot, and is not seen so distinctly as in some of the instances just mentioned ; but the great quantity of pudding-stone near it, renders it more interesting than it would be otherwise. As

---

* Theory of the Earth, Vol. I. p. 429.

the greater part of this little island is surrounded
by secondary strata, other junctions might be expected to be visible.

195. On the coast of England and Wales, from
the Solway Frith to the Land's End, though there
are several alternations from secondary to primary
strata, I know not that any of them have been observed. At St Bride's Bay, in Pembrokeshire, the
primary and secondary strata are seen very near
their junction ; but the precise line I believe is not
visible. The coal pits in the secondary strata, approach here within a few hundred yards of the primary. The secondary strata which commence at
this place, occupy both sides of the Bristol Channel, and meet the Cornish schistus, which extends
across the north of Devonshire to the Quantock
Hills, in a line that may be looked for on the
sea coast, somewhere between Watchett and Mine-
head.

196. Besides the sea coast, the beds of rivers
may be expected to afford information on this subject. To the instances I have mentioned, I have
accordingly two others from the inland country to
be added. One of them is from the river Jed, a
little way above Jedburgh, where the secondary
strata are seen lying horizontally on the primary,
a section of both being made by the bed of the
river. The phenomena here are very distinct, and
strongly marked : Dr Hutton has described and

represented them in a plate. *   He has mentioned another junction, not far from this, which he saw in the Tiviot.   Both these belong to the same primary ridge with the Siccar point.

197. I shall mention only one other, which was discovered by Lord Webb Seymour and myself, at the foot of the high mountain of Ingleborough, in Yorkshire.   As we went along the Askrig road from Ingleton, about a mile and a half from the latter, an opening appeared in the side of the hill, on the right, about one hundred yards from the road, formed by a large stone, which lay horizontally, and was supported by two others, standing upright.   On going up to the spot, we found it was the mouth of a small cave, the stone lying horizontally, being part of a limestone bed, and the two upright stones, vertical plates of a primary argillaceous schistus.   The limestone bed, which formed the roof of the cave, was nearly horizontal, declining to the south-east ; the schistus nearly vertical, stretching from north-west by west, to south-east by east.   The schistus, though close in contact with the limestone, seemed to contain nothing calcareous, and did not effervesce with acids in the slightest degree.

As this cave is at the foot of Ingleborough, a

---

* Theory of the Earth, Vol. I. p. 430 ; also Plate 3.

cold wind, 24° below the temperature of the external air, which issued from the mouth of it, might
very well be supposed to come from the inmost re-
cesses of that mountain.   Ingleborough, which
consists entirely of strata of limestone and grit,
nearly horizontal, and alternating with one another,
rises to the height of 1800 or 2000 feet above the
spot where we now stood.   This, I believe, is the
greatest thickness of secondary strata that has ever
been observed incumbent on the primary, and it is
therefore a geological fact highly deserving of attention.   The country all round, to a very great
extent, is composed of limestone, with a few beds
of grit interposed, and forming, beside Ingleborough, some other high mountains, such as Wharnside and Pennigant, all resting, it is probable, on
the same foundation.

At the spot just described, no breccia appeared
to be interposed between the primitive and secondary rock ; but we found a breccia at another point
of the same junction, not far distant.   This was at
a cascade, in the river Greta, called Thornton
Force, about two miles and a half from the place
just mentioned.   The Greta here precipitates itself from a horizontal rock of limestone ; and, after a fall of about eighteen or twenty feet, is received into a bason which it has worked out in the
primary schistus.   This schistus is in beds almost
perpendicular ; it exactly resembles that which has

just been described, and stretches nearly in the same direction. On the south side of the river a breccia was seen, lying upon the schistus, or rather, it might be said, that the lowest beds of limestone contained in them many rounded fragments of stone, which, on comparison, resembled exactly the schistus underneath. The primary rock itself is here seven or eight hundred feet above the level of the sea.

The same schistus, somewhat lower down the valley, and nearer to Ingleton, appears in large quantities, and is quarried for slate. Here, however, the immediate junction of the limestone and schistus does not appear.

I have dwelt longer on the description of these appearances than on any others of the same kind, because, from the great mass of secondary strata which here covers the primary, the circumstances are such as we cannot expect to see very often exemplified.

198. The Lakes of Cumberland are much visited by travellers; and it may be worth remarking, on that account, that, as the site of these lakes is a patch of primary country, bounded on all sides by secondary, so, in the rivers that run from the lakes, such junctions as we are now treating of may be expected to be found. Under Dun-Mallet, on the side toward Ulles Water, we observed a breccia, which was in horizontal layers, and seemed

to lie on the primary schistus, so that the whole hill is perhaps a piece of more indurated breccia, or secondary rock, which has resisted the wearing and washing down of the rivers better than the rest.

199. After ascertaining the fact of the disturbance of the strata, and their removal from their original position, it is of consequence to inquire into the direction of the force by which these changes have been produced. Now, if the disturbed or elevated strata, were every where in planes, without bending or sinuosity, it might perhaps be hard to determine, whether that force had acted in the direction of gravity, or in the opposite. Either supposition would account for the appearances ; and, as gravity is a known force, providing we can find some place fit to receive the matter impelled downward by it, its action would furnish the most probable solution of the difficulty.

It is on this principle that the Neptunian system proceeds, imagining, that certain great caverns or vacuities having been opened in the interior of the globe, a great part of the waters which formerly covered its surface, retired into them, and much of the solid rock also sunk down at the same time. In this way, one extremity of a stratum has been elevated, while the other has been depressed, and a certain inclination to the horizon has been given to the whole of it. Thus one cause serves two purposes; the vacuities in the interior of the earth account, both for the depression of the sea, and the elevation of the land; and the Neptunists, if the phenomena were all such as have been now stated, might boast of a felicity of explanation, not very usual in their system.

But this appearance of success vanishes, when the elevation and disturbance of the strata are more minutely examined, and are found to include waving and inflection, in a great variety of forms. It then becomes evident, that the beds of rock, at the time when they were disturbed from their horizontal position, had not their present hardness and rigidity, but were, in a certain degree at least, soft and flexible. Without these qualities, they could not have received, as they have often done, the curvature of a circle, not many feet, nay, not many inches, in diameter; nor could they have been bent into superficies, with their curvature in opposite directions, so that the same surface is in one part convex, and in another concave, on the same side, with a line of contrary flexure interposed. These are appearances, not reconcilable with the mere falling in, and breaking down of indurated rocks.

200. The inflections and wavings that we are here speaking of, though not peculiar to the primary strata, are found most frequently among them, and are perfectly familiar to every one who

has travelled among mountains with any view to
the study of geology.   The following are a few instances of this phenomenon out of a great number
which might be produced.

Saussure, in describing the route from Geneva
to Chamouni, mentions many remarkable instances of the bending of the strata, and particularly
where the small stream of Nant d'Arpenaz forms
a cascade, by falling over the face of a perpendicular limestone rock.   The strata of this rock are
bent into circular arches, extremely regular, and
with their concavity turned to the left.   What deserves particularly to be remarked, is, that a mountain behind the cascade has its strata bent in a direction opposite to the former, or with their concavity to the right.   There is no doubt that the
strata of both rocks are the same, so that a vertical
section of them would give a curve, in the figure
of an S. *   These circumstances are mentioned by
Saussure, and from them we may infer this other
property of these strata, that their section by a horizontal plane, must exhibit a system of straight
lines, probably all parallel to one another.

The same mineralogist describes the calcareous
strata which compose the mountain Achsenberg,
on the side of the Lake of Lucerne, as having from

---

* Voyages aux Alpes, Vol. I. § 472 ; also, Theory of the
Earth, Vol. II. p. 30.

top to bottom of the mountain the form of the letter S compressed, *(écrasée,)* with their curvature in some places very great. These inflections are repeated several times, and often in contrary directions; the layers are sometimes broken, where their curvature is greatest. *

On the side of the same lake, is another instance of bent strata, in a mountain, of which the beds are horizontal in the lower part, but are bent at one end upwards, in the form of the letter C. The horizontal part is of great extent, and the rock is also calcareous. †

The Montagne de la Tuile, near Montmelian, receives its name from the beds of rock being incurvated in form of a tile. ‡   Among secondary mountains, the same kind of phenomena are observed, though less frequently, and with less variety of inflection.   The chain of Jura is secondary, and the beds which compose it are of limestone, or of grit : they are bent in such a manner, that in a transverse section of the mountain, each layer would have the figure of a parabola. §

201. The Pyrenees furnish abundance of phenomena of the same kind, as we learn from the Essai

---

* Voyages aux Alpes, Tom. IV. § 1935.

† Ibid. § 1937.

‡ Ibid. Vol. III. § 1182, and Plate I.

§ Ibid. Tom. I. § 334.

sur la Minéralogie des Pyrenées. The calcareous strata of the valley of Aspe, represented Plate V. of that work, deserve particularly to be remarked.

202. Our own island abounds with examples of the bending and inflection of the strata, especially the primary, and many of them very much resembling those in the Alps and Pyrenees. On the top of the mountain of Ben-Lawers, in Perthshire, there is a rock, the face of which exhibits a section of a great number of thin equidistant layers, bent backwards and forwards like those described by Saussure ; and this unequivocal proof of the rock having once existed in the state of a flexible and tenacious paste, is rendered more striking, by the great elevation of the spot, and the ruggedness and induration, both of the stone itself, and of every thing that surrounds it. Many other mountains in this tract consist of a schistus, which is talcose rather than micaceous, and subject, in a remarkable degree, to the sort of sinuosity and inflection here treated of.

The appearances of the primary strata on the coast of Berwickshire, have been already mentioned, as affording much valuable instruction in geology. They also exemplify the waving and inflection of the strata on a large scale, and with great variety. A section of some of them is given by Dr Hutton, in his Theory of the Earth, Vol. I.

from a drawing made by Sir James Hall. The
nature of the curve superficies into which the schistus is bent, is the better understood from this,
that, besides transverse sections from north to
south, the deep indentures which the sea has made,
and the projecting points of rock, exhibit many
longitudinal sections, in a direction from east to
west.

203. The dock-yards at Plymouth are in several
places cut out of a solid rock of primary schistus,
singularly incurvated.   The inflections are seen
there to great advantage, being exhibited in three
sections, at right angles to one another, transverse,
longitudinal and horizontal.

204. From these instances, to which it were
easy to add many more, two conclusions may be
drawn.   The first of these is very obvious, viz.
that the strata must have been pliant and soft when
they acquired their present form.   The bending of
an indurated bed of stone into an arch of great
curvature, and without fracture, as in the preceding examples, is a physical impossibility.   Saussure
has indeed observed a fracture to accompany the
bending, in one or two cases ; but it is an uncommon phenomenon, and, where it happens, must no
doubt be understood to indicate an imperfect flexibility.   Now. if it be granted that the strata were
at any time soft and flexible, since their complete
formation, it will be found impossible to deny

their having been softened by the application of heat.

205. The second conclusion, alluded to above, results from a property, which belongs very generally, if not universally, to the inflections of the strata. This consists in their curvature being simple, or in one dimension only, like a cylindric superficies, not double, or in two dimensions, like the superficies of a sphere or spheroid. This may be otherwise expressed by saying, that the sections of the bent strata, by a horizontal plane, are straight lines, parallel to one another. On this account, every such stratum seems as if it were bent over an axis, and the axes of all these different bendings, for a great extent of country, are nearly parallel.

The truth of this is evident, where the strata are seen both transversely and longitudinally. It holds remarkably of the primary schistus on the coast of Berwickshire; where the beds of rock, if cut transversely, by a vertical plane, exhibit the figures of very complicated curves, with various *maxima* and *minima*, and points of contrary flexure; but, if they are cut by a horizontal plane, the section will produce nothing but straight lines, nearly parallel.

206. The constancy of the direction of the primary strata, when estimated by their intersection with the horizontal plane, is often very remarkable. Their elevation and flexure are subject to

great and sudden changes, so as to pass not only from greater to less, but from one side to the opposite, within a small distance ; but the horizontal line in which they *stretch*, usually preserves the same bearing to a great extent. The general direction of the primary strata, in the south part of Scotland, is from E.N.E. to W.S.W.; and the same is nearly true of those which compose the ridge of the Grampians on the north, and the hills of Cumberland and Westmoreland toward the south, though between the schistus of these three tracts, there is no communication at the surface, each being entirely separated from the one next it, by the interposition of secondary strata. I have already mentioned the observations of Lord Webb Seymour and myself, at the foot of Ingleborough ; and it appears from them, that the vertical schistus on which that mountain rests, though it still preserves an eastern and western direction, varies several points from that of the more northern strata. The strata of Wales return more to the first mentioned direction, and those of Devonshire and Cornwall agree with it very nearly. In all this, it will be easily conceived, that I do not mean to speak with absolute precision, or to deny the existence of great local irregularities. The result given is only a kind of average, deduced from observations hardly susceptible of great exactness, and not yet sufficiently multiplied to give to the conclusion all the accuracy it may attain.

207. This tendency of the primary strata to take a uniform direction, has also been observed in other countries. Saussure remarked in the Alps, that the beds of schistus are generally parallel to the chains of mountains composed of them ; * and this remark is probably applicable to all mountains consisting of primary strata. The general direction, therefore, of the schistus of the Alps, must be confined between W. 10° S. and W. 40° S. In the Pyrenees, the direction of the strata is about W.N.W. † If Saussure's rule may be depended on, the schistus of the Altaic, and most of the other great chains in the old continent, are in directions that run considerably to the south of west. The Urals, and perhaps some other of the northern chains, are however entirely different. In the Urals, as we learn not only from the general direction of the chain, but from a section of it in the 10th volume of the Nova Acta of Petersburgh, (Tab. 12,) the direction of the strata is nearly from N. to S. This last is probably the direction in the great chains of South America ; so that the uniformity of direction in the primary strata, which some mineralogists would extend to those of the

* Voyages aux Alpes, Tom. I. § 577.
† Essai sur la Mineralogie des Pyrenées.

whole earth, is certainly imaginary, though there
can be no doubt that it extends over very large
portions of the earth's surface. *

---

* It is perhaps unnecessary to observe, that the two propositions, that the intersections of the strata with the horizon
are parallel lines; and that they are lines which preserve the
same bearing with respect to the points of the compass; are
nearly the same thing for tracts of moderate extent, but for
large portions of the earth's surface are extremely different.
If, for instance, the belt of primary vertical schistus, which
traverses the south of Scotland, were to be produced eastward in the same plane, from its northern extremity, where
its direction is E.N.E. and its latitude 55° 57', it would cut
the meridian always less obliquely as it advanced, till, having
increased its longitude about 26° 28', it would be at right
angles to the meridian, and its direction of consequence due
east and west. This would happen in the parallel of 58° 51',
(on the shore of the Gulf of Finland, near Revel,) the strata
being now extended about 880 G. miles from the Siccar
Point. Conversely, vertical strata, having the same bearing
with respect to the meridian, may be in planes very much
inclined to one another. A stratum which bears east and
west in Cornwall, and one that does the same at the east end
of the Altaic chain, will be in planes, which, if produced,
would cut one another at right angles. All this is sufficiently
plain from the doctrine of the sphere, and is mentioned here,
merely as a caution to prevent too hasty conclusions from
being drawn from any correspondence of bearing among the
strata of remote countries.

For the sake of those who would deduce the medium
bearing of any body of strata from a number of observations,
it may be proper to take notice, that the true average is not

208. The tendency of the primary strata to remain straight in the horizontal direction, and to
be bent in the vertical, is a phenomenon which
points very directly to the causes from whence it
has arisen. A surface of simple curvature, or a
surface straight in one direction, is what the application of forces to different points of a plane, which
is flexible, though with a certain degree of rigidity,
will naturally produce. The supposition, therefore, that these strata were once flat and horizontal, and were impelled upward from that situation
before they had become rigid or hard, will explain
their having the kind of curvature which removes
them as little as possible from their original condition. But no other hypothesis affords any reason
why they should have that curvature more than any
other. From the falling in of roofs of caverns, we
might expect fracture and dislocation, without any
order or regularity; but certainly no bending or
sinuosity, nor any symmetrical arrangement. If,

---

to be found by simply taking an arithmetical mean among
all the observations. A more exact way is to work by the
traverse table, as in keeping a ship's reckoning, (supposing
the distance run to be always unity,) and to compute from
the observed bearings the amount of all the southing or
northing, and also of all the easting or westing. The sum of
all the latter, divided by the sum of all the former, is the
tangent of the angle which the general direction of the strata
makes with the meridian.

as some mineralogists allege, the curvature, as well
as inclination of the strata, arose from the irregularities of the bottom on which they were deposited, why is the former in one dimension only, and
why is it not in every direction, like that of hills
and valleys, or the actual surface of the earth ? Or,
lastly, if the whole structure of the primitive mountains is an effect of crystallization, and if these
mountains are now such as they have ever been
from the time of their consolidation, whence is it,
that, in their bendings the law just mentioned is
so constantly observed ? Indeed, the idea of a-
scribing the inflections of the strata to crystallization, though suggested by Saussure, * and since
become a favourite system with several mineralogists, appears to me in the highest degree unsatisfactory and illusive. The purpose for which crystallization is here introduced, is not to give a specific figure to a particular substance, but to arrange
the substances which it has formed and figured,
according to certain rules ; a work which we know
not how it is to perform, and in which we have no
experience of its power. Accordingly, this principle does not account, in any way whatever, for
the circumstances which attend the inflection of
the strata, for the simple curvature which they affect, nor for that parallelism of their layers, which,

---

* Voyages aux Alpes, Tom. I. § 475.

in all their bendings, is so accurately preserved. It does, indeed, so little serve to explain these facts, that, were the appearances completely reversed ; did the strata assume the most complex, instead of the most simple curvature ; instead of equidistant, were they converging, or alternately receding and approaching to one another ; the theory of crystallization might be equally applied to them. The state of the phenomena is a matter of perfect indifference to such a theory as this : all things are explained by it with the same facility ; the straight and the crooked, the square and the round, the moveable and the immoveable. Is it not evident that such an explanation is a mere word ; or, if any thing more than a word, an expression of our ignorance, so awkward and indirect, as to deprive us of whatever credit might have been gained by a plain and candid avowal of it?

It should never be forgotten, that a theory which accounts for *any thing*, and a theory which accounts for *nothing*, stand precisely on the same footing, and ought to be banished from all parts of philosophy, as they have been from those sciences which are justly honoured with the name of accurate. The animated orbs of Aristotle, and the vortices of Descartes, have long ceased to be mentioned in physical astronomy ; the first, because, they accounted for every thing alike ; the second, because, when they accounted for one thing, they never could be made to account for another. Both theories, therefore, have very properly been rejected ; and, when geology shall undergo a similar purification, the principle we have been considering will not be the only sacrifice required of the Neptunian system.

209. An appearance observed in some kinds of primary schistus, which clearly indicates their deposition by water, and in planes very different from those in which we now see them, though it might have been introduced before, is also much connected with the present argument. This appearance consists of small wavings or undulæ on the surface of the plates of schistus, precisely similar to those marks which are left by the sea on a gently inclining beach of sand, at the ebbing of the tide. All the species of schistus do not seem to afford instances of these wavings. The rocks which do so, are, I think, chiefly of the argillaceous kind, but often highly indurated ; so that the laminæ containing the impressions are not to be torn asunder but with great difficulty. Instances of it abound in the schistus of Berwickshire, and are also not unfrequent in that of Galloway. All must agree about the agent which produced these marks ; it could be no other than the sea ; but it must have been the sea acting on loose, small and round par-

ticles, lying on a surface which was nearly horizontal.

210. Dr Hutton's theory is no where stronger,
than in what relates to the elevation and inflection
of the strata ; points in which all others are so
egregiously defective. The phenomena to be connected are here extremely various, and even in appearance contradictory : the horizontality of one
part of the strata ; the inclined or vertical position
of another ; the perfect planes in which one set
are extended ; the breaking and dislocation found
in a second ; the inflection and sinuosity of a third ;
and almost every where the utmost rigidity and induration, combined with appearances of the greatest softness and flexibility ; the preservation of a
parallelism of superficies in the midst of so much
irregularity, and the assumption of a determinate
species of curvature, under circumstances the most
dissimilar ; all these appearances were to be connected with one another, and with the consolidation
of the strata, and this is done by the twofold hypothesis, of aqueous deposition, and the action of subterraneous heat. When these circumstances are
fairly considered, and when the shifts which other
systems are put to on this occasion are remembered, I think it will be granted, that few attempts at
generalization have been more successful, than that
which has been made by the Huttonian Theory.

211. To the fact of the elevation of the strata,

the study of geology is much indebted.   The stratified form of a great proportion of the earth's surface, gives to minerals that organization and regularity, which makes their disposition an object of
science, and their inclined position serves to bring
that organization into view, from far greater depths
than we can ever reach by artificial excavations.
If, for instance, the termination of strata, that make
with the horizon an angle of 30°, lying one over another, is seen for a horizontal distance of two miles ;
then it is certain, that if these strata have that extent under ground, which may be reasonably supposed, the thickness of the whole mass, measured by
a line perpendicular to its stratification, is half the
horizontal distance, or amounts to one mile.   It
would also require a pit to be sunk from the uppermost of these strata, to the depth of (2 miles $\times$
tan 30°, $=$) 6093 feet, before it could intersect the
undermost ; and therefore, if we suppose the same
stratum to preserve the same character for the extent of some miles, we obtain the same information
from inspecting the edge-seams, and see in reality
as far into the bowels of the earth, as if we had
sunk a perpendicular shaft to the depth of 6000
feet.

In general, the length of the horizontal line
drawn across the strata, from the lowest in position
to the highest, multiplied into the sine of the inclination of the strata to the horizon, gives the thick-

11

ness of the whole, measured perpendicularly to the plane of the stratification : and the same horizontal distance, multiplied into the tangent of the inclination, gives the actual depth at which the lowest stratum would meet a perpendicular to the horizon, drawn from the highest extremity of the upper stratum.

In many cases, the extent of stratified materials admitting of such an examination as this, is much greater than has now been supposed. M. Pallas describes a range of hills on the south-east side of the peninsula of the Tauride, which is cut down perpendicularly toward the sea, and offers a complete section of the parallel beds of a primary, or, as he calls it, an ancient limestone, inclined at an angle of 45° to the horizon ; and this section continues for the length of 130 *versts*, or about 86 English miles. The beds are so regular, that M. Pallas compares them to the leaves of a book. * The height of these hills does not exceed 1200 feet, but the real height of the uppermost stratum above the undermost, is $86 \times \sqrt{\frac{1}{2}} = 86 \times \frac{5}{7} = 61$ miles nearly.

If therefore we conceive that there is no shift in all this great system of strata, we in reality are enabled, by means of it, to see no less than 61 miles into the interior of the earth, nearly a 65th part of

* See Nova Acta Acad. Petropol. Tom. X (1792,) p. 257

the radius of the globe. It is true, that we can hardly suppose so great a body of strata to have been raised without shifting, so that we must diminish this depth considerably ; but were it reduced even to one-half, it will appear, that men see much farther into the interior of the globe than they are aware of, and that geologists are reproached without reason for forming theories of the earth, when all that they can do is but to make a few scratches on its surface. Art indeed can do little more ; but nature supplies the deficiency, and makes discoveries to the attentive observer, on the same great scale with her other operations.

The simplest account that can be given of the vast body of parallel and highly inclined strata just mentioned, is, that it consists of the ends of horizontal strata, or of strata not greatly inclined, that have been forced up when they were all soft and flexible. This is a much more conceivable supposition than Pallas's, viz. that the greater part of this mass has sunk down into some vast cavern in the interior of the earth.

NOTE XIII. § 53.

Metallic Veins.

212. The large specimens of native iron found in Siberia and Peru, mentioned above, § 51, are

among the most curious facts in the natural history
of metals. It has been doubted, however, by some,
whether they really belong to natural history, or
are not rather to be accounted artificial productions.
If they had been found in the heart of rocks, or in
the midst of metallic veins, no doubt of this sort
could possibly have been entertained ; but, as they
lie quite on the surface, in the middle of flat countries, and at a distance from any known vein of metal, the conjecture that they may be artificial, and
the remains of the iron founderies of ancient and
unknown nations, is at first sight not entirely destitute of probability. This probability, however,
will appear to be the less, the more carefully the
specimens are examined. The metal is too perfect,
and the masses too large, to have been melted in
the furnaces, or to have been transported by the
machinery, of a rude people. The specimen in
South America weighs 300 quintals, or about 15
tons, and is soft and malleable. * The Siberian
specimen, described by Pallas, is also very large ; it
is soft and malleable, and full of round cavities,
containing a substance, which, on examination, has
been found to be chrysolite. † Now, it is certainly quite impossible, that, in an artificial fusion, so
much chrysolite could have come by any means to be

---

* Phil. Trans. 1788, p. 37[; also p. 183, &c.
† Kirwan's Mineralogy, Vol. II. art. Native Iron.

involved in the iron ; but, if the fusion was natural, and happened in a mineral vein, the iron and the chrysolite were both in their native place, and their meeting together has nothing in it that is inexplicable.

213. Some circumstances in the description of the specimen in South America, such as the impressions of the feet of men and of birds on its surface, are not to be accounted for on any hypothesis, and certainly require more careful investigation. It is said, that this iron is very little subject to rust, and the analysis of a piece of it by Proust makes it probable, that it owes this quality to its union with nickel. * It appears, also, that the country of Chaco, where this specimen was found, affords many others of the same kind, one of which is mentioned in the description above referred to. That country lies on the east side of the Plata, and is a plain extremely level, and of vast extent, without any appearance of mineral veins ; but such veins may nevertheless exist undiscovered, in a tract subject to periodical inundations, and where the native rock is covered with alluvial earth and gravel to a great depth. The veins may be washed away, and the more durable substances, such as those pieces of native iron, may be left behind ; and, though they must be of a formation extremely ancient, ac-

* Annales de Chimie, Tom. XXXV. Messidor, p. 47.

cording to this hypothesis, they may not have been very long on the surface.

214. Specimens of native iron have been found, less remarkable than the preceding for their size, but in circumstances that excluded all idea of artificial fusion. Of this sort was Margraaf's specimen of native iron, the first of the kind that was known; it consisted of small bits of soft and malleable iron, found in the heart of a brown iron stone. * This makes it certain, that native iron is a natural production, and the mere circumstance of great magnitude, in the specimens before mentioned, does not entitle us to doubt of their having that same origin. It is a circumstance, besides, not in the least material to this argument; the smallest piece of native iron being as much a proof of fusion as the greatest; and the specimen of Margraaf being just as conclusive in favour of the Huttonian Theory, as those of Pallas or De Celis, supposing their reality as mineral productions to be completely established. A metal malleable and ductile, in ever so small a quantity, cannot be the result of precipitation from a menstruum, without a very particular combination of circumstances. Such a metal, on the other hand, can be readily produced by igneous fusion; so that here the negative and affirmative parts of the inductive argument may both be regarded as complete.

---

* Kirwan's Mineralogy, Vol. II. p. 156.

215. Mr Kirwan, in order to account for the magnitude of the two large specimens mentioned above, supposes, that small pieces of native iron (about the formation of which he appears to have no difficulty) have been originally agglutinated by petroleum, and left bare, when the surrounding stony or earthy masses either withered or were washed off. *   This is no doubt the most singular of all the opinions which have been advanced on the subject ; and, as it borrows nothing from analogy, it admits of no proof, and requires no refutation. None but a chemist of eminence could have ven‧ tured with impunity on an assertion so inconsistent with all the phenomena and principles of his science.

216. A remark of the same author, on the subject of the native gold found in the county of Wicklow in Ireland, is entitled to more attention. " That these lumps of native gold," he says, " were never in fusion, is evident from their low specific gravity, and the grains of sand found in the midst of them.   I found the specific gravity of a lump of the size of a nutmeg to be only 12800, whereas, after fusion, it became 18700." †

This argument is plausible ; but, I think, nevertheless inconclusive.   The sand found in the gold, accounts, at least in part, for its lightness. It is only by repeated fusions that any of the me-

---

* Geol. Essays, p. 405.   † Ibid. p. 402.

tals is brought to its utmost purity and highest specific gravity; and on no supposition can the melting of gold in the mineral regions, be very likely to separate it from heterogeneous substances. That quartzy sand should be found in it, after such a process, is naturally to be expected. The impressions which the quartz crystals have left on the Wicklow gold, would be received as a full proof of the fusion of that metal, if geologists always regulated their theories by the principles which determine the belief of ordinary men.

217. Don Rubin de Celis, in the paper referred to above, mentions some masses of silver found at Quantajaia, and also some dust of platina, in terms that excite a strong desire to have more information concerning them. They are considered by him as effects of volcanic fire; so we may conclude, that they contain evident marks of fusion, and would in this system be ascribed to that heat, from which volcanic fire is but a partial and accidental derivation.

218. The state also in which gold and silver are often found pervading masses of quartz, and shooting across them in every direction, furnishes a strong argument for the igneous origin, both of the metal and the stone. From such specimens, it is evident, that the quartz and the metal crystallized, or passed from a fluid to a solid state, at the same time ; and it is hardly less clear, that this

fluidity did not proceed from solution in any menstruum : For the menstruum, whether water or the
chaotic fluid, to enable it to dissolve the quartz,
must have had an alkaline impregnation; and, to
enable it to dissolve the metal, it must have had,
at the same time, an acid impregnation. But
these two opposite qualities could not reside in the
same subject; the acid and alkali would unite together, and, if equally powerful, form a neutral
salt, (like sea-salt,) incapable of acting either on
the metallic or the siliceous body. If the acid
was most powerful, the compound salt might act
on the metal, but not at all upon the quartz; and if
the alkali was most powerful, the compound might
act on the quartz, but not at all on the metal. In
no case, therefore, could it act on both at the same
time. Fire or heat, if sufficiently intense, is not
subject to this difficulty, as it could exercise its
force with equal effect on both bodies.

219. The simultaneous consolidation of the
quartz and the metal is indeed so highly improbable, that the Neptunists rather suppose, that the
ramifications in such specimens as are here alluded
to, have been produced by the metal diffusing itself through *rifts* already formed in the stone.
But it may be answered, that between the channels in which the metal pervades the quartz, and
the ordinary cracks or fissures in stones, there is no

---

* Geol. Essays, p. 401.

resemblance whatever: That a system of hollow tubes, winding through a stone, (as the tubes in question, must have been, according to this hypothesis, before they were filled by the metal,) is itself far more inconceivable than the thing which it is intended to explain; and lastly, that if the stone was perforated by such tubes, it would still be infinite to one that they did not all exactly join, or inosculate with one another.

220. The compenetration, as it may be called, of two heterogeneous substances, has here furnished a proof of their having been melted by fire. The inclusion of one heterogeneous substance within another, as happens among the spars and druses, found so commonly in mineral veins, often leads to a similar conclusion. Thus, from a specimen of chalcedony, including in it a piece of calcareous spar, Dr Hutton has derived a very ingenious and satisfactory proof, that these two substances were perfectly soft at the same time, and mutually affected each other at the moment of their concretion. *

Each of these substances has its peculiar form, which, when left to itself, it naturally assumes; the spar taking the form of rhombic crystals, and the chalcedony affecting a mammalated structure, or a superficies composed of spherical segments, contiguous to one another. Now, in the specimen under consideration, the spar is included in the

---

* Theory of the Earth, Vol. I. p. 93.

chalcedony, and the peculiar figure of each is impressed on the other ; the angles and planes of the spar
are indented into the chalcedony, and the spherical
segments of the chalcedony are imprinted on the
planes of the spar.   These appearances are consistent with no notion of consolidation that does not
involve in it the simultaneous concretion of the
whole mass ; and such concretion cannot arise from
precipitation from a solvent, but only from the congelation of a melted body.   This argument, it
must be remarked, is not grounded on a solitary
specimen, (though if it were it might still be perfectly conclusive,) but on a phenomenon of which
there are innumerable instances.

221. According to this theory, veins were filled
by the injection of fluid matter from below ; and
this account of them, which agrees so well with
the phenomena already described, is confirmed by
this, that nothing of the substances which fill the
veins is to be found any where at the surface.   It
is not with the veins as with the strata, where, in
the loose sand on the shore, and in the shells and
corals accumulated at the bottom of the sea, we
perceive the same materials of which these strata
are composed.   The same does not equally hold
of metallic veins : " Look," says Dr Hutton, " into
the sources of our mineral treasures ?   Ask the
miner from whence has come the metal in his
veins ?   Not from the earth or air above, nor
from the strata which the vein traverses : these do

not contain an atom of the minerals now considered. There is but one place from whence these
minerals may have come; this is the bowels of the
earth; the place of power and expansion; the
place from whence has proceeded that intense heat,
by which loose materials have been consolidated
into rocks, as well as that enormous force, by which
the regular strata have been broken and displaced." *

222. The above is a very just and natural reflection; but if, instead of interrogating the miner,
we consult the Neptunist, we will receive a very
different reply. As this philosopher never embar-
rasses himself about preserving a uniformity in the
course of nature, he will tell us, that though it
may be true, that neither the air, the upper part
of the earth's surface, nor even the sea, contain at
present any thing like the materials of the veins,
yet the time was when these materials were all
mingled together in the chaotic mass, and constituted one vast fluid, encompassing the earth; from
which fluid it was, that the minerals were precipitated and deposited in the clefts and fissures of the
strata.

223. It is alleged, in proof of this hypothesis,
that mineral veins are found to be less rich as they
go farther down, whereas they ought to be richer,

---

* Theory of the Earth, Vol. I. p. 130.

if they were filled by the projection of melted matter from below.   But the fact, that mines are less
rich as they descend farther, though it may hold
in some instances, is not general, and may therefore be supposed to arise from local causes, such as
are, in respect of us, accidental, and beyond the
limits to which our theories can be expected to
reach.   Thus the mines of Mexico and Peru are
said to be subject to the preceding rule ; but
in the mines of Derbyshire and Cornwall, the
very contrary is understood to take place.   Besides, what we are pleased to call the riches of
a mine, are riches relatively to us, and relatively to a distinction which nature does not recognise.   The spars and veinstones which are thrown
out in the rubbish of our mines, may be as pre-
cious in the eyes of nature, as conducive to the
great objects of her economy, and are certainly as
characteristic of mineral veins, as the ores of silver
or gold, to which we attach so great a value.   Unless the former are in smaller quantity, or less
highly crystallized at great than at small depths,
which I believe is not alleged, no conclusion can
be drawn from substances which occupy in general but a small proportion of any vein, and, in their
dissemination through it, do not seem to be always guided by the same law.

224. Again, if the veins were filled by deposition from above, we ought to discover in them such

horizontal stratification as is the effect of deposition from water, and we should perceive no marks
of the materials having been introduced with violence into their place.   The Neptunists cannot
object to the trial of their theory by these two
facts.

   As to the first, it is acknowledged, that there
is a certain regular disposition of the substances in
mineral veins, as stated § 59, but it is one which
has hardly any thing in common with the real
phenomena of stratification.   It consists in the
distribution of the principal substances in coats
parallel to the sides of the vein, each substance
forming a separate coat.   In a vein, for instance,
containing quartz, fluor, calcareous spar, lead, &c.
we might expect to find a lining of quartz crystals,
applied immediately to the walls of the mine, and
following exactly the irregularities of their surface ;
next, perhaps, a coat of fluor, then of calcareous
spar, and last of lead ore in the centre of the vein,
the same order being observed on the opposite
side.   These successive coats, it is material to remark, are not in planes, but in uneven surfaces, of
which the inequalities are evidently determined by
those of the walls, that is, of the rock which forms
the sides of the vein ; neither are they horizontal,
but are parallel to the walls, whether these be perpendicular or inclined.   Here, therefore, there is
no appearance of the action of that statical law

which has directed the arrangement of the other strata, and which tends to make the plane of every stratum deposited by water perpendicular to the direction of gravity. The coating of the veins has therefore been performed under the conduct of some other power than that which presides over aqueous deposition. If, as the Neptunists maintain, the materials in the veins were deposited by water, in the most perfect tranquillity, it is wonderful that we do not find those materials disposed in horizontal layers, across the vein, instead of being parallel to its sides ; and it seems very unaccountable, that the common strata, deposited as we are told while the water was in a state of great agitation, have so rigorously obeyed the laws of hydrostatics, (§ 38,) and acquired a parallelism in the planes of their stratification, which approaches so often to geometrical precision ; while the materials of the veins, in circumstances so much more favourable for doing the same, have done nearly the reverse, and taken a position, often at right angles to that which hydrostatical principles require. This is a paradox which the Neptunian system has created, and which therefore it is not very likely to resolve.

225. Mere words should have little power to mislead, in a science which treats of sensible objects, such as are always easily subjected to the examination of sight or of touch ; yet there is some

10

appearance as if the Neptunists were misled in this, and other instances, by the term *stratification.* Though an incrustation on the perpendicular face of a rock has very little affinity to a stratum, such as we are accustomed to see deposited by water, yet the same name being once imposed on both, mineralogists have proceeded to reason concerning them, as if they were precisely the same thing, and were both to be ascribed to the same cause. Indeed, every perpendicular or highly inclined bed of stone, is inexplicable as an effect of aqueous deposition, in a system, unprovided, as the Neptunian is, * with the means of raising up such beds from a horizontal into a vertical position. This observation may also be extended to all cases of vertical stratification. Water cannot directly arrange its deposits in planes highly inclined, and therefore I have often wondered to see the Neptunists contending so eagerly for the stratification of certain rocks, such as granite, which, being vertical, or highly inclined, was much less friendly to their system than the entire absence of all stratification would have been. I was disposed to admire their candour, when the use which they made of the fact convinced me, that I ought only to wonder at their inconsequential reasoning. The Huttonian Theory is, indeed, the only one which possesses

---

* See preceding note.

the means of reconciling the elevation of the strata
with their horizontal deposition, and which is entitled to consider stratification, in whatever plane it
may be, as originally the work of the ocean.   The
geologists who attach themselves exclusively to the
action of water, will never be able to extend the
dominion of that element so far as Dr Hutton has
done, by combining it with fire.

226. But, though the Neptunian system were
provided with engines, powerful enough to raise
up strata from a level to a vertical plane, this would
avail nothing in the present instance ; since, on no
supposition, can the incrustations on the perpendicular sides of a vein have ever been horizontal.
On no supposition, therefore, can these incrustations be received as a proof of aqueous deposition :
it may indeed be certainly inferred from them, that
the matter which they consist of was fluid at the
time of their formation ; but the absence of all appearance of a horizontal disposition, in any part of
the vein, amounts nearly to a demonstration, that
this fluidity did not proceed from solution in a
menstruum. We must therefore conceive the coats
to have been formed during the refrigeration of the
melted matter injected from the mineral regions
into the clefts and fissures of the strata. (§ 59.)

227. Mineral veins, particularly at their intersections with one another, contain abundant marks
of the most violent and repeated disturbance, (§ 56.)

Not to mention that they owe their first formation to the fracture and displacing of rocks already consolidated, it appears, that they have originated at very different periods, and that the birth of each has been accompanied with convulsions, which shook the foundations of the earth. In Cornwall, for instance, the principal veins, and those which they distinguish particularly by the name of *Lodes*, have nearly the same direction with the strata or vertical schistus, extending from about E. N. E. to W. S. W. These, however, are often intersected nearly at right angles by other mineral veins, called *Cross Courses*, and this hardly ever happens without the latter moving, or, as it is called, *heaving* the former out of their direction. This plainly indicates, that the cross courses are of later origin than the others, and that their formation was accompanied with such a force, as must, in many instances, have moved the whole body of rock which constitutes the promontory of Cornwall, and probably much more, for several yards, in a horizontal direction. Sometimes, also, both the longitudinal and the cross vein are forced out of their place by a third. These disturbances arise not only from mineral veins, but from veins of porphyry and granite, the production of which has been attended with no less violence than of the others.

228. What is here said of Cornwall, is the history, in some degree, of all mineral countries what-

ever.   The great horizontal *translation* which has thus accompanied the formation of veins; the movement impressed on such vast bodies of rock, and the frequent renewal of these immense convulsions; are not to be explained by the mild and tranquil dominion of the watery element.   They require the utmost power that is known any where to exist, and were it not for the admonitions of the volcano and the earthquake, we might doubt if even subterraneous heat itself possessed an energy adequate to these astonishing effects.

229. From the *heaving* of one vein by another, it is evident, that there was a force of protrusion in the direction of one of them, that acted at the time of its formation.   This force cannot be accounted for on the supposition that veins were produced by the mere shrinking of the strata; for the rocks could not, in that case, have been rent asunder, and impelled forward at the same time.   It appears most likely, that fissures in the strata were made, at least in many instances, and the matter poured into them, nearly at the same time, both being effects of the same cause, the expansive force of subterraneous heat.

230. It is remarked, at § 56, that the shifting of the strata is best observed where the veins make a transverse section of beds of rock, considerably inclined to the horizon.   It is also true, that in some cases the near approach of the strata to the

level, may make the shifts produced by the veins
very easy to be discovered.   Thus in Derbyshire,
where the mineral veins are in secondary strata,
nearly horizontal, there is almost no instance in
which the corresponding strata are not observed to
be on different levels, on the opposite sides of the
same vein.

231. The fact described by Deluc, and referred
to at § 55, may, for what we know of it, admit of
being explained in two ways.   The great wedge of
rock which appears to be insulated between two
branches of the same vein, may either be a mass
that has been broken off, and sustained by the
melted matter that flowed all around it ; or, it may
be a mass of rock contained between two veins that
are in reality distinct, and of different formation.
Whether this last supposition is the truth, would
probably be evident from a careful examination of
both parts of the vein ; as some difference of character cannot fail to be the consequence of different
formation.   If no such difference is observed, the
two branches must be supposed to belong to the
same vein, and the only probable explanation of
the insulation of so large a mass of rock will be by
the first mentioned supposition.   This fact, therefore, notwithstanding the great attention M. Deluc has bestowed on it, still requires further examination, before it can be decided whether it inclines
to the Huttonian Theory, as on the first supposi-

tion, or is, as on the latter hypothesis, equally balanced between it and the Wernerian.

232. Whatever be the case with this fact, the
general one of pieces of rock being found insulated
in veins, is certainly favourable to the notion of an
injected and ponderous fluid having originally sustained them.   Where, as happens in some instances, the stones contained in the veins have no affinity to any of the rocks above, they cannot be supposed to have come any how but from below, and
to have been carried up by the matter of the vein.
The instance from the slip at the Huddersfield Canal has been already mentioned.

233. The preceding observations have been
principally directed against that theory of veins
which supposes them to have been filled by deposition from water.   There is another theory maintained by some of the Neptunists, that the metals
in veins were introduced there by infiltration. *
This opinion is sufficiently refuted by the fact, that
rarely any metallic ore is found out of the vein, or
in the rock on either side of it, and least of all
where the vein is richest.   This is inconsistent
with the notion of the ore being carried into the
vein by water percolating through the adjacent
rocks, unless some satisfactory reason is assigned,
which determined the water to leave the ore in

* Geol. Essays, p. 401.

the vein and no where else. Besides, this hypothesis does not account for the formation of the
spars and veinstones which fill the vein, and which
appear clearly to have been brought there at the
same time with the ore, and no doubt by the same
cause.

234. The veins, properly so called, are indefinitely extended ; but there are also thin plates of
spar, and of crystals of different kinds, often found
included in rocks, and shut in on all sides, to which
the name of veins is commonly applied. These
last ought certainly to be distinguished from the
former, and may not improperly be called Plate
Veins or Lenticular Veins, the plate or cake of
spar of which they consist having very often the
form of a lens, though, as may be supposed, considerably irregular. Either of these terms being derived entirely from external characters, has the advantage of involving nothing theoretical.

The lenticular veins are certainly not formed
like the usual mineral veins, by injection, since
they are shut in, on all sides, by the solid rock.
When they are found, therefore, in stratified rocks,
such as have not themselves been melted, we must
conceive them to be composed of materials more
fusible than the surrounding rock, so that they
have been brought into fusion by a degree of heat
which the rest of the rock was able to resist, and,
on cooling, have assumed a sparry structure. When

they are found in rocks, of which the whole has been fluid, they must be considered as component parts of that mass, which, by an elective attraction, have united with one another, and separated themselves from the substances to which they had less affinity.

The veins of this kind seem to be connected with those called in Derbyshire Pipe Veins, in which the ores of metals are sometimes found. The pipe veins, indeed, are not in all cases completely insulated, but sometimes communicate with the veins properly called mineral. I am too little acquainted, however, with their natural history, to be able to say with certainty to which of the two species they ought to be referred.

NOTE XIV. § 75.

On Whinstone.

235. To the facts and reasonings given above, I shall, in this note, add a few remarks, tending to show, that whinstone is not of volcanic, nor of aqueous, but certainly of igneous origin.

It is asserted, (§ 62,) that carbonate of lime and zeolite are often contained in whinstone, but never in lava, and that this circumstance may sometimes serve to distinguish these stones from one another.

With respect to carbonate of lime, in particular, it seems evident, that this substance cannot enter into the original composition of any lava, because the same heat which melted the lava, would, where there was no greater pressure than the weight of the atmosphere, expel the carbonic acid and produce quicklime. Notwithstanding this, rocks containing carbonate of lime, have often been considered as lavas, into the pores and cavities of which, calcareous matter having been carried by the infiltration of water, had crystallized into spar. Thus Spallanzani, in his account of the Euganean Hills, in Lombardy, describes some of the rocks as abounding at their surface, and even in their interior, with air-bubbles of various sizes, from such as are hardly perceptible, to some that are half an inch in diameter; and which, he says, are all of an oval figure, with their longest diameters in the same direction. This he considers as a proof that the rock is a genuine lava; for the air-bubbles prove the stone to have had its fluidity from fire; and by their elongation in the same direction they prove, that the mass when fluid was also in motion. Spallanzani adds, that *many of these cavities are filled with crystals of the carbonate of lime, an effect of the infiltration of water.* *

---

* Voyages dans les deux Siciles, Tom. III. p. 157. Edit. de Faujas de St Fond.

236. Though the argument here advanced for the igneous origin of the rock may be admitted as conclusive, the introduction of calcareous spar into it by infiltration must still be questioned. Lava, except in a state of decay or decomposition, is not readily penetrated by water; and, if it were, the filling of cavities with spar, by means of the water percolating through them, would still be subject to many difficulties. (§ 12.) Besides, whinstone rocks are frequently found so full of calcareous spar, or of zeolite, that they would become porous to such a degree, if the cavities filled with these latter substances were all empty, that they could hardly sustain their own weight, and much less that of the great masses of rock incumbent on them. In such cases, it is certain, that the crystallized substances were part of the original composition of the rock. The truth is, that the infiltration of the water is a mere gratuitous assumption, introduced for the purpose of explaining the existence of carbonated lime in a stone which had endured the action of intense heat : and this assumption ought of course to be rejected, if the phenomenon can be explained by a theory, that is in other respects conformable to nature. The spar, then, may be considered as a proof, that the rocks in question are to be numbered with those unerupted lavas which have flowed deep in the bowels of the earth, and under a great compressing force. This is the more probable,

8

that the Euganean Hills, like some whinstone hills
in our own country, have, in certain places, a
covering of slaty and calcareous strata incumbent on
them, even at their summits, * so that the torrent
of melted stone, of which they are admitted to
consist, cannot have flowed from the mouth of a
volcano. I do not mean to say, that there are
among these hills no vestiges of volcanic explosion.
1 am very far from having *data* sufficient for drawing this conclusion; but I believe it may be safely
affirmed, that the bulk of them is no more composed of volcanic lava, than the basaltes of Staffa,
or of the Giant's Causeway.

237. But, besides the evidence deduced from
calcareous spar and zeolite, against the rocks containing them being real lava, there are other marks,
even less equivocal perhaps, that distinguish the
lavas which we suppose to have flowed in the mineral regions, from those which have actually flowed on the surface. These are what we collect from
the disposition, the organization, or, as we may say,
the physical geography of whinstone countries, unlike, in so many respects, to that of volcanic countries. The shape of whinstone hills; their large
flat terraces, rising one above another; their perpendicular faces, and the correspondence of their
heights even at considerable distances; have no-

---

* Phil. Trans. 1775, p. 34.

thing similar to them in the irregular torrents of volcanic lavas. The phenomena of the former are also on a scale of magnitude very far exceeding the latter, and clearly indicate, that though both have been produced by fire, it has been by fire in very different circumstances, and regulated by very different laws. The structure of the two kinds of rock agrees, in many respects, and so does their chemical analysis; but their disposition and arrangement are so dissimilar, that they cannot be supposed to be of the same formation.

238. This argument, I believe, was first stated by Mr Strange, in a letter to Sir John Pringle, published in the 65th volume of the Philosophical Transactions. * That intelligent observer, after visiting the countries in Europe most remarkable either for burning, or for what are accounted, extinguished volcanoes, and examining them with a very discriminating eye, remained convinced, that there are two distinct species of rock, which both owe their origin to fire; but to fire acting in circumstances and situations extremely different. The first is the common volcanic lava; the other, to which he gives the name of a basaltine rock, comprehends such rocks as the Giant's Causeway, the

---

* Account of Two Giants' Causeways in the Venetian State, &c. by John Strange, Esq. Phil. Trans. Vol. LXV. (1775,) p. 5, &c.

l

basaltes of the Vivarais, of the Euganean Hills,
&c. and differs in nothing from that which is called
here by the name of whinstone.   Mr Strange conceived, that the one of these kinds of stone could,
no more than the other, be accounted the work of
aqueous deposition, but was led to the distinction
just mentioned, by observing the organization and
arrangement in the rocks of the latter kind, and
comparing them with the disorder and ruin that
every where mark the footsteps of volcanic fire.
He does not pretend to determine the nature of
the fire to which the basaltine rocks owe their formation, nor the circumstances in which it has acted : he is satisfied with the negative conclusion,
that it is not volcanic ; and his paper affords a specimen of what is perhaps rare in any of the sciences,
and certainly most rare of all in geology, viz. a philosophic induction carried just as far as the facts
will bear it out, and not a single step beyond that
point.

239. Several other hints contained in this paper
are highly deserving of notice ; for we not only
find in it the notion of a formation of basaltic rocks,
igneous though not volcanic, but also that of their
simultaneous crystallization, * together with the
suggestion, that granite and basalt are of the same
origin. †   These opinions had not, I believe, oc-

---

* Phil. Trans. *ubi supra*, p. 17.
† Ibid. p. 36 and 37.

curred at that time to any mineralogist except Dr Hutton, nor had they been communicated by him to any but a few of his most intimate friends ; so that Mr Strange has without doubt all the merit of a first discoverer. Indeed, without the knowledge of the principle of compression, such as it is laid down by Dr Hutton, it was hardly possible for him to proceed further than he has done. He remarked the *unburnt* limestone that lies on the tops of some of the Euganean basaltes, and seems to have been aware of the great difficulty, which it was reserved for the Huttonian Theory to overcome. His letter contains also some excellent general remarks on the rocks of the Vivarais and Velay, which he had visited, before Faujas de St Fond had published his curious and elaborate description of these countries.

240. The cause of the peculiar structure which has just been observed to distinguish whinstone from volcanic countries, is easily assigned in the Huttonian Theory. According to that theory, the whinstone rocks were formed, in the bowels of the earth, of melted matter poured into the rents and openings of the strata. They were cast, therefore, in those openings, as in a mould ; and received the impression and character of the rocks by which they were surrounded. Hence the tabular masses of whinstone, which when soft have been interposed between strata, and compressed by

their weight, so as almost to have themselves acquired the appearance of stratification. Hence the perpendicular faces of the same rocks, produced by their being abutted when yet soft, against the abrupt sides of the strata. The rocks which formed those moulds have, in many cases, entirely disappeared ; in others, a part still remains, surrounding, or even covering, the basaltes, as in the Euganean Hills, in those of the Val di Noto in Sicily, the rocks near Lisbon, * and in different parts of Great Britain.

Above all, the veins of whinstone which intersect the strata, are the completest proofs of the theory here given of these rocks, and the most inconsistent, in all respects, with the hypothesis of their volcanic origin.

241. If these *criteria* are applied to what are called extinguished volcanoes, I have no doubt that many which have been reckoned of that number, will be found to derive their origin more directly from the fire of the mineral regions. The basaltic rocks of the Vivarais, I am well persuaded, belong to this class; and I conclude that they do so, not only from the account of them given by Mr Strange, but from the description of Faujas himself, who, though under the influence of the

---

* Recherches sur les Volcans Eteints du Vivarais ; Lettre de Dolomieu, p. 443.

opposite theory, seems very fair and accurate in his description of phenomena.   The most unequivocal mark of real whinstone rock, and of a formation in the strictest sense mineral, is where veins of that kind of rock intersect the strata.   Now, in a letter to Buffon, on the streams of lava found in the interior of certain calcareous rocks in the lower Vivarais, Faujas describes what can be accounted nothing else but a vein or dike of whinstone, accompanied with several of its most remarkable and characteristic appearances : " Figurez-vous un courant de lave, de la nature du basalte noir, dur et compacte, qui a percé à travers les masses calcaires, et s'est fait jour dans quelques parties, paroissant et disparoissant alternativement : Cette coulée de matière volcanique s'enfonce sous une partie de la ville, bâtie sur le rocher ; elle reparoit dans la cave d'un maréchal, se cache et se montre encore de temps en temps en descendant dans le vallon, &c.   Ce qu'il y a d'admirable, c'est que la lave forme deux branches bien extraordinaires, dont l'une s'éleve sur la crête du rocher, tandis que l'autre coupe horizontalement de grands bancs calcaires escarpés, qui sont à découvert, et bordent le chemin.

" Quels efforts n'-a-t-il pas fallu pour forcer cette lave se prendre une telle direction, et se percer cette suite de rochers calcaires ?   Si cette longue coulée de lave avoit eu 200 ou 300 toises de largeur, je ne serois pas surpris qu'un torrent de matière en fusion de ce volume eut pu produire des effets extraordinaires et violens ; *mais figurez-vous, Monsieur, que dans les endroits les plus larges, elle n'a tout-au-plus qu'environ* 12 *ou* 15 *pieds ; elle n'en a que* 3 *ou* 4 *dans certaines parties.*" *

This narrow stream is to be traced across the strata for more than a league and a half ; and the whole appeared to Faujas so marvellous, that he says he almost doubted the testimony of his senses. He would have done much better, however, to have doubted the conclusions of his theory ; for it was by them that the phenomena before him were rendered so mysterious and incredible. While he continued to regard what is described above as a stream of melted lava, which had descended from the top of one mountain, and climbed up the sides of the opposite, like water in a conduit pipe, piercing occasionally through vast bodies of solid rock, it is no wonder that he considered as marvellous what is indeed physically impossible. Had his belief in the volcanic theory permitted him to see in all this, not a superficial current, but one of indefinite depth, he would have beheld the object divested, not of what was curious and interesting, but of what was incredible or absurd, and reduced to the

---

* Volcans Eteints du Vivarais, p. 328, &c.

VOL. I.   S

same class of things with mineral veins. That it belongs really to this class, and is no more than a vein or dike of whinstone, intersecting the strata to an unknown depth, and most probably, like other veins, communicating with the mineral regions, cannot be doubted by any one who has studied the subject of basaltine rocks, through any other medium than the volcanic theory. The ramifications which run from it into the calcareous rock, contrived, Faujas says, just as if on purpose to perplex mineralogists, is one of the well known and characteristic appearances of basaltic veins.

242. It can hardly be doubted, that the lava described by the same author as heaving up a mass of granite, * and including pieces of it, is a rock of real whinstone. The same may be said of many others ; and, though I pretend not to affirm that there is nothing volcanic in the Vivarais, I must say, that nothing decidedly volcanic appears in the description of that country, but many things that are certainly of a very different origin.

In the present state of geological. science, a skilful mineralogist could hardly employ himself better, than in traversing those ambiguous countries, where so much has been ascribed to the ancient operation of volcanic fire, and marking out what belongs either clearly to the erupted or unerupted

---

* Volcans Eteints du Vivarais, fol. p. 365, &c.

lavas, and what parts are of doubtful formation, containing no mark by which they may be referred to the one of these any more than the other. Such a work would contribute very materially to illustrate the natural history of the earth.

243. One of the most ingenious attempts to support the volcanic theory, is the system of *submarine volcanoes*, imagined by the celebrated mineralogist Dolomieu. The phenomenon that led to this hypothesis, was what he had observed in the hills near Lisbon, and still more remarkably in those of the Val di Noto in Sicily, where the basaltine rocks had regular strata incumbent on them, and in some cases interposed or alternated with them. * It seemed from this evident, that the strata were of later formation than the stone on which they rested ; and as they must, on every supposition, be held to be deposited by water, it was concluded, that the lava which they covered had been thrown out by volcanoes at the bottom of the sea ; that the strata had afterwards been deposited on this lava ; and that, in some cases, there had been frequent alternations of these eruptions and depositions. †

---

* Mémoire de Deodate de Dolomieu, sur les Volcans Eteints du Val di Noto, en Sicile. Journal de Phys. Tom. XXV. (1784, Septembre,) p. 191.

† Near Vizini, in the Val di Noto, Dolomieu tells us, that he counted eleven beds, alternately calcareous and volcanic, in the perpendicular face of a hill, which at a distance appear-

244. Though this hypothesis does certainly deliver the system of the Volcanists from one great difficulty, it is itself liable to insurmountable objections. I shall just mention some of the principal.

1. The regular and equidistant strata that we often see covering the tops of whinstone or basaltic rocks, could not have been deposited in the oblique and very much inclined position which they now occupy.

This is remarkable in the strata which cover the basaltic rock of Salisbury Crag, near Edinburgh, at its northern extremity. The strata are very regular, and must have been deposited in a plane nearly horizontal ; yet the surface of the basaltes on which they now rest is very much inclined, dipping rapidly to the north-east. The necessity of a horizontal deposition in strata, which, though not now horizontal, have their planes nearly parallel to one another, has been proved at § 38.

2. If there is any truth in the principles established above, even the strata themselves have not been consolidated without the action of fire. By Dolomieu's system, therefore, the consolidation of the strata which cover the basaltes is not accounted for.

---

ed like a piece of cloth, striped black and white; *ubi supra.* In another instance he saw more than twenty of these alternations. He has since made similar observations in the Vicentine and in Tyrol. Journal de Phys. Tom. XXXVII. (1790,) Partie 2, p. 200.

3. There are no means furnished by the hypothesis of submarine volcanoes for bringing the basalt, and the strata which cover it, above the level of the sea. If it is said that the waters of the sea have been drained off, the objections are all incurred that have been stated at § 37.* If it is said, that the rocks themselves have been elevated by a force, impelling them upwards, we say, that the existence of such a force, when admitted, furnishes another means of explaining the whole phenomenon, namely, that of the injection of melted matter among the strata, the same that is used in the Huttonian Theory.

4. The phenomena of basaltic veins are not in the least explained by the hypothesis of submarine volcanoes. That hypothesis, then, even if the foregoing objections were removed, does not serve to explain all the facts respecting the rocks of this genus, and wants, of consequence, one of the most important characters of a true theory. It must be allowed, however, that it makes a considerable approach to such a theory, and that the submarine volcanoes of Dolomieu, have an affinity to the unerupted lavas of Dr Hutton.

245. Though in these remarks I have endeavoured to expose the errors of the volcanic system,

---

* Dolomieu adopts this supposition; he thinks that the surface of the sea must have been formerly 500 or 600 toises above its present level. Ibid. p. 196.

I cannot but consider that system as coming infinitely nearer to the truth than the Neptunian.
It has the merit of distinguishing an order of rocks,
which bears no mark of aqueous formation, and
in which the crystallized, sparry, or lava-like structure, bespeaks their primeval fluidity, and refers
their origin to fire. The Neptunian system, on the
other hand, strives to confound the most marked
distinction in the mineral kingdom, and to explain
the formation, both of the stratified and unstratified rocks, by the operation of the same element.
Though chargeable with this inconsistency, it has
become the prevailing system of geology; and the
arguments which support it are therefore entitled
to attention.

246. It will no doubt be thought singular, that
the same mineralogist, whom we have just seen
exerting his ingenuity in defence of the volcanic
system, should now appear equally strenuous in defence of the Neptunian. Though Dolomieu contends for the volcanic origin of some basaltic rocks,
he does not admit that all basaltes is volcanic, nor
even all of igneous formation. Thus he states, that
he had examined at Rome some of the most ancient monuments of art, executed in basaltes,
brought from Upper Egypt, and that he could discover no mark of the action of fire in any of them.*

---

* Journal de Physique, Tome XXXVII. (1790,) Partie 2,
p. 193.

On the contrary, he found that some of them consisted of green basaltes, which changes its colour to a bronze, when exposed even to a moderate heat, and which therefore, he argues, can never have endured any strong action of fire.

The answer to this argument is very plain, if we admit the effects ascribed by Dr Hutton to the compression which necessarily takes place in the mineral regions. If indeed the heat in those regions resembled exactly that of our fires at the surface, it would not be easy to deny the above conclusion, which therefore certainly holds good against the volcanic origin of the Egyptian basaltes. But there is no reason why, under strong compression, the colouring matter of these stones might not be fixed, and indestructible by heat, though it can be easily volatilized or consumed when such compression is removed. This argument then is against the volcanic; but not against what has been called the *Plutonic* formation of basaltes.

247. As to the other marks of fire which Dolomieu sought for and did not find in the above-mentioned stones, we are not exactly informed in what they consisted. If the crystallized or spathose texture that belongs to this description of stones was wanting, the specimens were not to be considered as of the real basaltic or whinstone genus, whatever their name or history may seem to indicate. If they did possess that texture, they had

the only mark of an igneous origin that could be expected, supposing that origin to have been in the bowels of the earth. No part, therefore, of the observations of this ingenious mineralogist, can be considered as inconsistent with the theory of basaltic rocks which has been laid down above. ‾

248. Bergman had before reasoned on this subject precisely in the same manner, but from better data, as the stones from which he derived his argument were in their native place : " Trap," says that ingenious author, (that is whinstone,) " is found in the stratified mountains of West Gothland, in a way that deserves to be described. The lower stratum, which is several Swedish miles in circuit, (10½ of these miles make a degree,) is an arenaceous stone, horizontal, resting on granite, and having its particles agglutinated by clay. The stratum above this is calcareous, full of the petrifactions of marine animals, and above this is the trap. These three kinds of rock compose the greater part of the mountains just mentioned, though there are some other beds, particularly very thin beds of marl and of clay, which separate the middle stratum, both from that which is under it and over it, and are frequently so penetrated with bitumen that they burn in the fire. This schistus is black ; when burnt it becomes red, and afterwards, when washed with water, affords alum. How can it be supposed," he adds, " that the

trap has ever been violently heated, while the
schistus on which it is incumbent retains its black-
ness, which however it loses by the action even of
a very weak fire ?" *

The answer to this argument is already given.
The reasoning, as in the former instance, is conclusive only against the action of volcanic fire, or
fire at the surface; but not against the action of
heat deep in the bowels of the earth, and under
the pressure of the superincumbent ocean. In such
a situation, the bituminous schistus might be in
contact with the melted basalt, and yet there might
be no evaporation of the volatile, nor combustion
of the inflammable parts. It does not, however,
always happen, that the bituminous substances, or
substances alterable by fire, which are found in
contact with basaltes, are without any mark of having endured the operation of fire. Instances in
which such operation is apparent are given above,
§ 30; and more will be added in the conclusion of
this note.

249. The same mineralogist founds another argument for the aqueous formation of whin or trap,
on the existence of that stone in the form of veins,
included in primeval rocks: " Invenitur hoc
saxum (trap) in Suecia pluribus locis, sæpeque in

---

* Bergman de Productis Volcanicis, Opuscula, Tom. III.
p. 214, &c.

montibus primævis, angustas implens venas, adeo subtilis structuræ, ut particulæ sint impalpabiles, et, dum niger est, genuinum efficit lapidem Lydium. In hisce montibus, nulla adsunt ignis subterranei vestigia." *

The phenomena here described, namely, a vein of compact whinstone traversing a primary rock, is, without doubt, as incapable of being explained by the operation of a volcano, as it is by that of aqueous deposition. It is, however, a most complete proof of the original softness of the substance of which the veins consist, and affords one of the strongest possible arguments for such an operation of fire as is supposed in the present theory. The main arguments, therefore, which have been proposed as subversive of the igneous origin of basaltes, are only subversive of their formation by one modification of fire, viz. of fire acting near the surface ; and thus the weapons which directly pierce the armour of the Volcanist, and inflict a mortal wound, are easily turned aside by the superior temper of the Plutonic mail.

250. An argument founded on facts very similar to some of the preceding, and leading to the same conclusion, is employed by the mineralogist to whom the Neptunian system owes its chief support. Werner, in his observations on volcanic

---

* Opuscula, ubi supra.

rocks and on basaltes, has rested his proof of the
aqueous formation of the latter, on their interposition between beds of stone in mountains regularly
stratified, and obviously formed by water. He describes an instance of this in the basaltic hill of
Scheibenberg ; and the facts, though most of them
are not uncommon, are highly deserving of attention. Near the top of this hill, and above the basaltic rock which composes the body of it, he tells
us, that there was a sand-pit ; a circumstance which
he appears to consider as not a little singular. It
was, however, at the bottom of the hill, that he
met with the appearances which chiefly attracted
his notice : " First," says he, " or lowest, was a
thick bank of quartzy sand, above that a bed of clay,
then a bed of the argillaceous stone called wacke,
and upon this last rested the basaltes." " When
I saw," adds he, " the three first beds running
almost horizontally under the basaltes, and forming its base ; the sand becoming finer above, then
argillaceous, and at last changing into real clay, as
the argil was converted into wacke in the superior
part ; and, lastly, the wacke into basaltes ; in a
word, when I found a perfect transition from pure
sand to argillaceous sand, from the latter to a sandy
clay, and from this sandy clay, through many gradations, to a fat clay, to wacke, and at last basaltes,
I was irresistibly led to conclude, that the basaltes,
the wacke, the clay, and the sand, are all of one

and the same formation ; and that they are all the effect of a chemical precipitation during one and the same submersion of this country." *

First, as to the sand on the top of this basaltic hill, it is most probably the remains of certain sand-stone strata that originally covered the basaltic part, but are now worn away.   We are therefore

---

* " Combien je fus surpris de voir en arrivant au fond, un épais *banc de sable quartzeux,* puis au-dessus une *couche d'argile,* enfin une couche de la pierre argileuse nommée *wacke,* et sur celle-ci reposer le *basalte.*   Quand je vis les trois premières couches s'enfoncer *presqu' horizontalement sous le basalte,* et former ainsi sa *base ;* le sable devenir plus fin au-dessus, puis argileux, et se changer enfin en vraie argile, comme l'argile se convertissoit en wacke dans sa partie supé‑rieure ; et finalement la wacke en basalte ; en un mot, de trouver ici une *transition parfaite* du *sable pur* au *sable argil‑leux,* de celui-ci à *l'argile sablonneuse,* et de *l'argile sablon‑neuse,* par plusieurs gradations, à *l'argile grasse,* à la *wacke,* et enfin au *basalte.*

" A cette vue, je fus sur le champ et irrésistiblement én‑trainé à penser, (comme l'auroit été sans doute tout connois‑seur impartial frappé des conséquences de ce phénomène ;) je fus, dis-je, irrésistiblement entrainé aux idées suivantes : Ce *basalte,* cette *wacke,* cette *argile,* et ce *sable, sont d'une seule et même formation ;* ils sont tous l'effet d'une *précipita‑tion par voie humide* dans une seule et même submersion de cette contrée ; les eaux qui la couvroient alors transportoient d'abord le *sable,* puis deposoient *l'argile,* et changoient peu‑à-peu leur précipitation en *wacke,* et enfin en vraie *basalte.*"— Journal de Physique, Tom. XXXVIII. (1791,) Partie 1, p. 415.

to consider this as an instance of a basaltic rock, interposed between strata that are undoubtedly of marine origin. In this, however, there is nothing inconsistent with Dr Hutton's theory of basaltes; on the contrary, it is one of the principal facts on which that theory is founded. It has indeed been argued by some mineralogists, that bodies thus contiguous must owe their origin to the same element, and that a mineral substance cannot be of more recent formation than that which lies above it. But the maxim, that a fossil must have the same origin with those that surround it, does not hold, unless they have a certain similarity of structure. It is, for instance, the want of this similarity, that authorizes us to assign different periods of formation to mineral veins, and to the rocks in which they are included.

In a succession of strata, no one can doubt, that the lowest were the first formed, and the others in the order in which they lie; but, when between two strata of sandstone or of limestone we find an intermediate rock, so different as to resemble lava, and to have nothing schistose or stratified in its composition, the same instrument cannot be supposed to have been employed in the formation of both; nor is there any reason why we may not suppose, that the intermediate body was interposed between the other two, by some action subsequent to their formation. It was thus that Dolomieu concluded, when he saw a lava-like stone interposed between calcareous strata in the Val di Noto, that, though contiguous, these two rocks could not possibly be of the same formation ; and thus far it is certain, that every unprejudiced observer must agree with him.

251. But the circumstance on which Werner seems to lay the greatest stress, is the gradual transition from the sand to the basalt, through the intermediate steps of clay and wacke ; this gradual transition he considers as a direct proof, that they are all of the same formation.

A gradual transition of one body into another, can only be said to take place, when it is impossible to define their common boundary, or to determine the line where the one begins and the other ends. Now, if this be the proper notion of gradual transition, I must say, that after much careful examination, I have never seen an instance, in which such a transition takes place between whinstone and the contiguous strata. The *line* of separation, though in some places less evident than in others, has, on the whole, been marked out with great precision ; and, though the stones have been firmly united, or, as one may say, welded one upon another, yet, when a fresh fracture was obtained, the stratified and unstratified parts have rarely failed to be distinguished. The fresh fracture is indeed often necessary, for many species of whinstone get by decomposition a granulated texture at the surface, so as hardly to be distinguished from real sandstone.

Some of the kinds of primary schistus also, particularly the argillaceous, when much indurated, have in their structure a considerable resemblance to whinstone ; they are slightly granular, or laminated, and have a tendency to a sparry texture. Where it happens that this sort of schistus and whinstone are contiguous, it is natural to expect, that their common boundary will be traced with difficulty, and in many parts will be quite uncertain.   Still, however, if a careful examination is made ; if the effects of accidental causes are removed ; and, above all, if the more ambiguous instances are compared with the more decisive, and interpreted by them, though single specimens may be doubtful, we will hardly ever find that any uncertainty remains with respect to entire rocks.

252. This general fact, which I state on much better authority than that of my own observations, viz. on those of Dr Hutton, is not given as absolutely without exception.   The theory of whinstone which has been laid down here, leads us indeed to look for some such exceptions.   It is certain, that the basis of whinstone, or the material out of which it is prepared by the action of subterraneous heat, is clay in some state or other, and probably in that of argillaceous schistus.   It fol-

lows, of consequence, that argillaceous schistus may by heat be converted into whinstone. When, therefore, melted whinstone has been poured over a rock of such schistus, it may, by its heat, have converted a part of that rock into a stone similar to itself; and thus may now seem to be united, by an insensible gradation, with the stratum on which it is incumbent; and phenomena of this kind may be expected to have really happened, though but rarely, as a particular combination of circumstances seems necessary to produce them. Hence it is evident, that stones may graduate into one another, without being of the same formation; and that it is fallacious to conclude, from the insensible transition of one kind of rock into another, without any other circumstance of affinity, that they have both the same origin.

I am disposed, therefore, to make some limitation to what is said in § 72, where I have expressed an absolute incredulity as to such transitions as are here referred to. The great skill and experience of the mineralogist who has described the strata at Scheibenberg, do not allow us to doubt of his exactness, though some of the appearances are such as decomposition and wearing might well enough be supposed to produce. The fairest way is to take Werner's observations just as they are given us, and to try whether they cannot be explained without the assistance of his theory. In effect, the wacke which he describes, rests, it would seem, on an unconsolidated bed of clay; and it may be supposed, that a part of this bed has been converted into wacke by the heat of the incumbent mass, and has thus produced the apparent gradation from the one substance to the other. As the appearances of the rocks of Scheibenberg seem to be considered by Werner as furnishing a very strong, and even an unexpected confirmation of his system, I cannot help thinking, that an explanation of them, on the principles of Dr Hutton, without any straining or forcing of those principles, contributes not a little toward extending the empire of the latter over all the phenomena of geology.

253. Another fact, which has been much insisted on of late, in proof of the aqueous formation of basaltic rocks, is that shells are found in them. Of the reality of this fact, however, or at least of the instances hitherto produced, great doubts I think may be reasonably entertained. The specimens of the supposed basaltes, with shells included in them, that are chiefly relied on, are found at Portrush in Ireland, a rocky promontory to the westward of the Giant's Causeway, and separated from it by a considerable body of calcareous strata. Some of these specimens were brought to Edinburgh about a year ago, and were supposed, I believe, to contain an irrefragable proof of the Nep-

tunian origin of the basaltic promontory where they were found. I went to see these specimens in company with Lord Webb Seymour and Sir James Hall ; and, on examining them carefully, we were all of opinion, that the stones which contained the shells, or the impressions of the shells, were no part of the real basaltes. They were all very compact, and had all more or less of a siliceous appearance, such as that of chert ; they had nothing of a sparry or crystallized structure ; their fracture was conchoidal, and but slightly uneven. In two of them, one of which bore the impression of a *cornu ammonis,* the schistose texture might be distinctly perceived. A specimen which accompanied them, but in which there was no shell, served very exactly to explain the relation between these stones and the true basaltes. Part of this specimen was a true basalt, and the rest a sort of hornstone, exactly the same with that in which the shells were, and not unlike the jasper that is under the whinstone of Salisbury Crag, and in contact with it ; so that on the whole it was evident, that the rock containing the shells is the schistus or stratified stone, which serves as the base of the basaltes, and which has acquired a high degree of induration, by the vicinity of the great ignited mass of whinstone.

This solution of the difficulty has since been confirmed by observations made on the spot by Dr

Hope, who discovered two or three alternations of the basaltic rock, with the beds of the schistus in which the shells are contained.

254. This also explains some observations of Spallanzani, made in the island of Cerigo, on the coast of Greece, the Cythæra of the ancients.* The base of that island is limestone ; but it abounds also in unstratified rocks, which the Italian naturalist supposes to be of volcanic origin ; but which, if I mistake not, we would regard as whinstone, or perhaps porphyry ; and they are said to contain oyster shells and pectinites of a large size, perfectly mineralized. These petrifactions, however, Spallanzani says, are not contained in the lava that has actually flowed, but in stones which have only endured a slighter action of fire. Without the commentary afforded by the Portrush specimens, it would be difficult to make out any thing very precise from this description. By help of the information derived from those specimens, we may conclude, that the condition of the shells in them, and in the rocks of Cerigo, is perfectly alike ; and that, in both cases, the shells are involved in parts of the rock which are truly stratified, but which have been, in some degree, assimilated to the basaltes by the heat which they have endured. Spallanzani would probably have used exactly the same terms which he

---

* Journal de Physique, Tom. XLVIII. (1798,) p. 278.

employs in speaking of Cerigo, if he had been required to describe the petrified shells at Portrush.

255. In the instances just mentioned, the petrified marine objects are not found in the real whinstone ; but if they were found in it, when it borders
on stratified rocks containing such objects, the thing
would not be at all surprising, nor furnish any argument against the igneous consolidation of the
stone. If a torrent of melted matter was poured in
among the strata, by a force which at the same time
broke up and disordered those strata, nothing could
be more natural, than that this matter should contain fragments of them, and of the objects peculiar
to them.

In one instance, mentioned by Mr Strange, this
seems actually to have taken place.   In the Vero-
nese, a country remarkable for a mixture of limestone strata, containing marine objects, with volcanic or basaltine hills, he assures us, that he had seen
a mass of stone, which had evidently concreted from
fusion, in which the marine fossil bodies, originally, as he supposes, contained in the strata, were
perfectly distinguishable, though variously disfigur-
ed. *   It may be, that in this, as in the foregoing
examples, it was not real basaltes, or real lava,
which contained the shells, but the conterminal
rock ; but, supposing it to be as Mr Strange repre-

---

* Phil. Trans. 1775, p. 25.

sents it, there appears to be no inconsistency between the phenomenon, and the igneous origin of
the rock in which the shells were included. Here,
however, it should be remarked, that the presence
of great pressure, to prevent the conversion of the
shells into quicklime, seems absolutely necessary;
and that the phenomenon of these basaltic petrifactions, requires the application of heat to have been
deep under the surface of the earth.

256. The phenomena we have been considering,
have been selected as the most unfavourable to the
igneous origin of basaltic rocks ; and we have seen,
that when duly examined, they are not at all inconsistent with it. We are now to take a view of
some appearances, that seem quite irreconcilable
with the aqueous formation of these rocks.

Where whinstone rocks are found in masses,
bounded by the strata, and insulated among them,
they subject the Neptunian system to great difficulties. For, supposing it true that this stone may be
produced by the precipitation and crystallization of
mineral substances dissolved in water, yet it seems
unaccountable, that this effect has been so local and
limited in extent, as often to be confined to an irregular figure of a few acres, while, all round, the
substances deposited have had no tendency to crystallization, and have been formed into the common
secondary strata. The rock of Salisbury Crag, for
instance, is a mass of whinstone, having a perpen-

dicular face eighty or ninety feet high toward the west, and extending from north to south with a circular sweep about 900 yards. The whole of this rock rests on regular beds of secondary sandstone, not horizontal, but considerably depressed toward the north-east : the rock is loftiest in the middle, and decreases in thickness toward each end, terminating at its northern extremity in a kind of wedge. It is covered at top, toward that extremity, with regular beds of sandstone, perfectly similar to those on which it is incumbent ; and it is not improbable, that this covering formerly extended over the whole.

Now, what cause can have determined the column of water, which rested on the base at present occupied by this rock, to deposit nothing but the materials of whinstone, while the water on the south, west, and north, was depositing the materials of arenaceous and marly strata ? Wherefore, within this small space, was the precipitate every where *chemical*, to use the language of Werner, while close to it, on either side, it was entirely *mechanical?* Why is there, in this case, no gradation ? and why is a mere mathematical line the boundary between regions where such different laws have prevailed ? Whence also, we may ask, has the basaltic deposit been abruptly terminated toward the west, so as to produce the steep face which has just been mentioned ? The operation of currents, or of any motion that can take place in a fluid, will furnish no explanation whatever of these phenomena; yet they are phenomena far from being peculiar to a single hill; they are among the most general and characteristic appearances in the natural history of whinstone mountains; and a geological theory which does not account for them, is hardly entitled to any consideration.

257. The basaltic rock, just described, is also covered, at least partly, with strata perfectly similar to those that lie under it. Now, it appears altogether unaccountable, that after the water had done depositing the materials of the whin on the spot in question, the former order was so quickly resumed, and a deposition of sand, and of the other materials of the strata, took place just as before. All this is quite unintelligible; and the principles of the Neptunian system seem here to stand as much in need of explanation, as any of the appearances which they are intended to account for.

258. The unequal thickness, and great irregularity in the surface of the whinstone mass, here treated of, and of many rocks of the same kind, is also a great objection to the notion of their aqueous formation. This seems to have been perceived by Werner, in the instance of the rocks formerly mentioned; and he endeavours to explain it, by sup-

posing, that much of these rocks has been destroyed by waste and decomposition, so that an irregularity of their surface, and want of correspondence
has been given to them, which they did not originally possess.   In the instance of Salisbury Crag,
however, we have a proof, that the great irregularity of surface, and the inequality of thickness, do
not always arise from these causes.   The thinnest
part of that rock, toward its northern extremity,
is still covered by the strata in their natural place,
and has been perfectly defended by them from
every sort of wearing and decay.   The cuneiform
shape, therefore, which this rock takes at its extremities, and the great difference of its thickness at
them and in the middle, is a part of its original
constitution, and can be attributed to nothing casual, or subsequent to its consolidation.

The same may be said of many other basaltic
rocks, where an inequality of thickness, most unlike to what belongs to aqueous deposits, is known
to exist in beds of whinstone that are still deep under the surface.   Thus the toadstone of Derbyshire, even where it has a thick covering of strata
over it, has been found, by the sinking of perpendicular shafts, to vary from the thickness of eighteen
yards to more than sixty, within the horizontal distance of less than a furlong.   Nothing of this
kind is ever found to take place in those beds of

rock which are certainly known to originate from aqueous deposition, and no character can more strongly mark an essential difference of formation.

259. We have had frequent occasion to consider the characters of those masses of whinstone which are so often found interposed between stratified rocks. These have been found in general very adverse to the Neptunian system; and two of them which yet remain to be mentioned, are even more so than any of the rest.

Where a bed or tabular mass of whinstone is interposed between strata, and wherever an opportunity offers of seeing its termination, if the strata under it are not broken, it may be remarked, that they do not abut themselves bluff and abrupt against the whin. On the contrary, if we mark the course of the stratum which covers the whinstone, and of that which is the base of it, we shall find they converge toward one another, the interposed mass growing thinner and thinner, like a wedge. When the latter terminates, the two former come in contact, and have no stratum interposed between them. Thus the roof and base of the whinstone rock are contiguous beds, that appear as if they had been lifted up and bent, and separated by an interposed mass. Had the whole been an effect of simultaneous deposition, the regular strata must have been abruptly terminated by the whin, like two courses of different sorts of masonry where they meet with one another.

260. From this wedge-form of the whinstone masses, and in general from the irregularity of their surfaces, another conclusion follows, similar to the preceding, and one which has been already mentioned. Where the surface of the interposed mass is greatly inclined to the horizon, the strata which rest on this inclined plane, are nevertheless as exactly parallel to that plane, and to one another, as if they were really horizontal. It is certain, therefore, that they were not deposited on the same inclined plane on which they now rest; for, if so, they would have been still nearly horizontal, and by no means parallel to the inclined side of the whinstone. This follows from the nature of aqueous deposition, as already explained.

We have a remarkable instance of the phenomenon here referred to, in the rock of Salisbury Crag, of which mention has been so often made, and in which almost every circumstance is united, that can serve to elucidate the natural history of basaltic rocks. The north end of that rock is in the figure of a wedge, with its inclined side considerably steep, and covered by strata of grit, perfectly regular, and parallel to the surface on which they lie. The inspection of them will convince any one, that they were not deposited by the water, on a bottom so highly inclined as that on which they now rest. They are of a structure very schistose ; their layers very thin ; so that any inaccuracy of their parallelism would be readily observed. The appearances of the horizontal deposition of these strata, are indeed so clear, and so impossible to be misunderstood, that the followers of the Huttonian system would not risk much, if they were to leave the whole theory of whinstone to the decision of this single fact, and should agree to abandon that theory altogether, if the Neptunists can show any physical or statical principle, on which the deposition now described can possibly have been made, or will point out the rule, by which nature has given a structure so nicely stratified to arenaceous beds deposited on a surface so highly inclined. If no such principle can be pointed out, though we cannot conclude that the Huttonian Theory is true, we certainly may conclude that the Neptunian is false.

261. Proofs of the igneous formation of whinstone, still more direct, are derived from the induration of the contiguous strata ; from their disturbance when intersected by veins of whinstone ; and from the charring of the coal which happens to be in contact with these veins. These are considered above at § 66, 67, &c. ; and it is particularly taken notice of at § 66, that pieces of sandstone are sometimes found as if floating in the whinstone, and, at the same time, greatly altered

in their texture. One of the best and most unequivocal instances of this sort which I have seen, is
to be found on the south side of *Arthur's Seat*,
near Edinburgh. The rock which composes the
upper part of the hill, on that side, is a whinstone
breccia, such as we have many examples of, and, I
believe, very much resembling what is called a *lava
brecciata* by the volcanic geologists. The stony
fragments included in this compound mass, are for
the greater part rounded ; and some of them are of
whinstone, others of porphyry, strongly characterized by rectangular maculæ of feldspar, and many
seem to be of sandstone, but so considerably altered, as to leave it at least disputable whether they
really are so or not. In one part, however, where
the face of the rock is nearly perpendicular, a narrow ridge is seen standing out from the rest, and
of a different colour, being more entirely covered
with moss than the rock round about it, and, as
may be presumed from that circumstance, less liable
to decomposition. On examination I found, that
this ridge does not consist of whinstone, but of a
very hard and highly consolidated sandstone. It
appears to be the edge of a stratum, of the thickness of about nine or ten inches, and of the height
of fifteen or sixteen feet. It is not perfectly straight,
but slightly waved, its general direction being nearly vertical ; and it is on both sides firmly embraced
by the whinstone. When broken, it appears that

this sandstone resembles in colour, and in every thing but its greater consolidation, and more vitreous structure, the common grit found at the bottom of the hill, and over all the adjacent plain.

262. If all these circumstances are put together, there appears but one conclusion that can be drawn from them. We have here the manifest marks of some power which could lift up this fragment of rock from its native place, distant at least several hundred yards from its present situation, place it upright on its edge, encompass it with a solid rock, of a nature quite heterogeneous to itself, and bestow on it, at the same time, a great addition of solidity and induration. If the mass in which this stone is now imbedded, be supposed to have been once in fusion, and forcibly thrown up from below, invading the strata, and carrying the fragments along with it, the whole phenomena now described admit of an explanation, and all the circumstances accord perfectly with one another ; but, without this supposition, they are so many separate prodigies, which have no connection with one another, nor with any thing that is known. It is indeed impossible, that the effects of motion and heat can be more clearly expressed than they are here, or the subject in which these powers resided more distinctly pointed out.

263. The preceding facts being susceptible but of one interpretation, are on that account extreme-

ly valuable. The phenomena of Salisbury Crag, near the same place, are almost equally free from ambiguity. The basaltic rock which forms that precipice, rests on arenaceous or marly strata ; and these, in their immediate contact with the former, afford an instance of what is mentioned § 67, namely, the conversion of the strata in such situations into a kind of petrosilex, or even jasper. The line which separates the one rock from the other, is, at the same time, so well defined, as, in the eyes even of the most determined Neptunist, to exclude all idea of insensible gradation.

264. The same rock affords some remarkable instances of the disturbance of the strata contiguous to the whinstone. The beds of the former are bent upwards in several places ; and, at one in particular, form an arch, with its convexity downward, so as to make it evident, that the force which produced this bending was directed from below upwards.

265. It is, however, where whinstone takes the form of veins, intersecting the strata, that the induration of the latter is most conspicuous. The coast of Ayrshire, and the opposite coast of Arran, exhibit these veins in astonishing variety and abundance. The strata are, in many instances, so *reticulated* by the veins, and intersected at such small distances, that it seems necessary to suppose, that the fissures in them were hardly sooner made than

10

filled up. This at least is true, if the veins are to
be accounted all of the same formation ; and, in the
greatest number of instances by far, there is no
mark of the one being posterior to the other.

266. The induration of the sides of these veins,
in some cases, has been such, that the sides have
become more durable than the vein itself; so that
the whinstone has been worn away by the washing
of the waves, and has left the sides standing up,
with an empty space, like a *ditch*, between them.
One of these I remarked on the south side of Bro-
dick Bay, in Arran, which, where it met the face of
an abrupt cliff, was not less than forty or fifty feet
in depth.

267. I shall pass over whatever argument might
be drawn in favour of our system, from the slender
ramifications of the veins, and the varieties of their
sizes, from a few inches to many fathoms in diameter, and also from the connection which they often
appear to have with the great tabular masses of basaltes ; and shall only add a few remarks on the
charring of coal in the vicinity of veins or masses
of whinstone. The connection between the charring of coal and the presence of whinstone, was first
observed by Dr Hutton ; and, as far as opportuni-
ties of verifying the observation have yet occurred,
appears to be a fact no less general than it is curious and interesting. In the coal mines of Scotland, it certainly holds remarkably, particularly in

those about Saltcoats in Ayrshire, where a whinstone dike is known to stretch across the whole of
the coal country, and to be every where accompanied
with blind or uninflammable coal. At Newcastle,
dikes of the same kind are met with, and one, in
particular, in what is called the *Walker* Colliery,
has proved the action of subterraneous fire, to the
satisfaction of mineralogists nowise prejudiced in
favour of the Huttonian system.

The coal found under basaltes, in the Island of
Sky, has been already mentioned, § 140. To what
was said concerning the fibrous structure of the
parts of that fossil in immediate contact with the
whin, it may be added, that it is also charred in
those parts, so as to have hardly any flame when it
is burnt, though further down it is of the nature of
ordinary coal. Indeed, if there be any truth in Mr
Kirwan's general remark, that it is common to find
wood coal under basaltes, it must be understood to
arise from this, that the coal in contact with the
basaltes is frequently charred, and its fibrous structure, by that means, rendered more visible.

268. It has been objected to the supposition of
coal having its bituminous part driven off by the
heat of the whinstone, that this ought not, on Dr
Hutton's principles, to happen in the mineral regions. But it may be replied, as has been done
above, that the local application of heat might certainly produce this effect, and might drive off the

11

volatile parts from a hotter to a colder part of the same stratum. The bitumen has not been so volatilized and expanded as entirely to escape from the mineral regions ; but it has been expelled from some parts of a mass, only to be condensed and concentrated in others. This supposition coincides exactly with the appearances.

269. The native or fossil coke which accompanies whinstone, has been distinguished into two varieties. The first is the most common, in which, though the coal is perfectly charred, it is solid, and breaks with a smooth and shining surface. The second is also perfect charcoal, but is very porous and spungy. This substance is much rarer than the other. Dr Hutton mentions an instance of it at the mouth of the river Ayr, where there is a whinstone dike. * I had the satisfaction of visiting it along with him. It was in the bed of the river, below the high water mark ; the specimens had the exact appearance of a *cinder*.

In the banks of the same river, some miles higher up, he found a piece of coal, belonging to a regular stratum, involved in whinstone, and extremely incombustible. It consumed very slowly in the fire, and deflagrated with nitre like plumbago. This he considered as the same fossil which has been described under the name of *plombagine*.

---

* Theory of the Earth, Vol. I. p. 611.

Near it, and connected with the same vein of whinstone, was a real and undoubted plumbago.

From these circumstances he also concluded, that plumbago is the extreme of a gradation, of which fossil coal is the beginning, and is nothing else than this last reduced to perfect charcoal. This agrees with the chemical analysis, which shows plumbago to be composed of carbon, combined with iron.

In confirmation of this theory, he mentions a specimen, in his possession, of steatitical whinstone, from Cumberland, containing nodules of a very perfect and beautiful plumbago ; and he also takes notice of a mine of this last, in Ayrshire, which, on the authority of Dr Kennedy, who has examined it with great care, I can state as being contained, or enveloped in whinstone ; and I hope the public will soon be favoured with a particular description of this very interesting spot, by the same ingenious and accurate observer.

270. Thus the mineralogical and chemical discoveries agree in representing coal, blind coal, plombagine, plumbago, as all modifications of the same substance, and as exhibiting the same principle, carbon, in a state of greater or less combination.   As the last and highest term of this series should be placed the *diamond ;* but we are yet unacquainted with the matrix of this curious fossil, and its geological relation to other minerals.

When known, they will probably give to this substance the same place in the geological, as in the
chemical arrangement: in the mean time, it is
hardly necessary to remark, how well all the preceding facts agree with the hypothesis of the igneous formation of whinstone, and how anomalous
and unconnected they appear, according to every
other theory.

271. Notwithstanding all this accumulated and
unanswerable evidence for the igneous formation
of basaltes, a great objection would still remain to
our theory, were it not for the very accurate and
conclusive experiments concerning the fusion of
this fossil, referred to above, § 75.   A strong prejudice against the production of any thing like a
real stone by means of fusion, had arisen, even
among those mineralogists, who were every day
witnesses of the stony appearance assumed by volcanic lava.   They still maintained, on the authority of their own imperfect experiments, that nothing but glass can ever be obtained by the melting
of earths or of stones, in whatever manner they are
combined.

An ingenious naturalist, after describing a
block of basaltes, in which he discovered such appearances, as inclined him to admit its igneous
consolidation, rejects that hypothesis, merely from
the imaginary inability of fire to give to any substance a stony character: " Quelque mélange,"

says he, " de terres que l'on suppose, quel que soit
le degré de feu que l'on imagine, quel que soit le
tems que l'on emploie, il est très certain que l'on
n'obtiendra pas, par le seul fluide igné, ni basalte,
ni rien qui lui ressemble." *

Sir James Hall's experiments have completely
demonstrated the contrary of what is here asserted : they have added much to the evidence of the
Huttonian system ; and, independently of all theory, have narrowed the circle of prejudice and
error.

NOTE xv. § 83.

On Granite.

1. *Granite Veins.*

272. It is said above, § 77, that granite is found
in unstratified masses, and in veins.   In the former of these conditions, it constitutes entire mountains, and forms the central ridge of many of the
greatest chains that traverse the surface of the
earth.   It is the granite of this kind that has been
most generally described by travellers and mineralogists.   The veins have not been so much attend-

* Journal de Phys. Tom. XLIX. (1799,) p. 36.

ed to, though they are of peculiar importance for ascertaining the relation between granite and other fossils.

273. Though Dr Hutton was the first geologist who explained the nature of granite veins, and who observed with attention the phenomena which accompany them, he is not the first who has mentioned them. M. Besson found veins of this kind in the Limoges, in an argillaceous schistus, and unconnected, as far as appeared, with any large mass of granite. *

Saussure met with granite veins in the Valorsine, but did not see them distinctly. He ascribed them to infiltration. † The date of this observation is in 1776 : He afterwards discovered similar appearances at Lyons. ‡

Werner also, in enumerating the substances of which veins are formed, reckons granite as one of them.

274. Veins of granite may be considered as of two kinds, according as they are connected, or not connected apparently with any large mass of granite. It is probable, that these two kinds of veins only differ in appearance, and that both are connected with masses of the same rock, though that connection is visible in some instances, and invisible in

---

* Journal de Phys. Tom. XXIX. p. 89.
† Voyages aux Alpes, Tom. I. § 598, 599.
‡ Ibid. § 601.

others. The distinction, however, whatever it be
with respect to the thing observed, is real with
respect to the observer; and, as it is right, in a
description of facts, to avoid every thing hypothetical, I shall speak of these veins separately.

275. Veins of granite, having no communication, so far as can be discovered, with any mass of
the same rock, are found in the Western Islands of
Scotland, particularly in that of Coll, where they
traverse the beds of gneiss and hornblende schistus,
which compose the main body of the island. They
are sometimes several fathoms in thickness, obliquely intersecting the planes of the strata just
mentioned, which are nearly vertical. In these
veins the feldspar is predominant; it is very highly crystallized, and of a beautiful flesh colour.
Many smaller veins are also to be met with in the
same place; but no large mass of granite is found,
either in this or the adjacent island of Tiree.

276. The Portsoy granite, of which mention
has been already made, § 80, also constitutes a
vein or dike, traversing a highly indurated micaceous schistus, about a mile to the eastward of the
little town of Portsoy, and not visibly connected
with any large mass of the same kind. More dikes
than one of this granite have been observed near the
same spot.

A similar granite is likewise found inland, in the
neighbourhood of Huntly, about eighteen miles

south of Portsoy ; but whether in the shape of a
vein or a mass, I have not been able to learn.

277. Veins of granite are also frequent in Cornwall, where they are known by the name of *lodes*,
the same name which is applied in that country to
metallic veins.   The granite veins frequently inter-
tersect the metallic, and are remarkable for producing shifts in them, or for throwing them out of
their natural direction.   The mineral veins, particularly those that yield copper and tin, run nearly
from east to west, having the same direction with
the beds of the rock itself, which is a very hard
schistus.   The granite lodes, as also those of porphyry, called *elvan* in Cornwall, are at right angles
nearly to the former ; and it is remarked, that
they generally heave the mineral veins, but that the
mineral veins seldom or never heave the cross
veins.   In this country, therefore, the veins of
granite and porphyry are posterior in formation to
the metallic veins.   These veins of granite may
perhaps be connected with the great granitic mass
that runs longitudinally through Cornwall, from
Dartmoor to the Land's End.   This much is certain, that their directions in general are such, that,
if produced, they would intersect that mass, nearly
at right angles.

278. The granite veins in Glentilt, where Dr
Hutton made his first observations on this subject,
are not, I believe, visibly connected with any large

mass of the same rock. * The bed of the river
Tilt, in the distance of little more than a mile, is
intersected by no less than six very powerful veins
of granite, all of them accompanied with such marks
of disorder and confusion in the strata, as indicate
very strongly the violence with which the granite
was here introduced into its place. These veins
very probably belong to the great mass of granite
which is known to form the central ridge of the
Grampians further to the north ; but they are several miles distant from it, and the connection is
perhaps invisible in the present state of the earth's
surface.

279. The second kind of granite vein, is one
which proceeds visibly from a mass of that rock,
and penetrates into the contiguous strata.   The
importance of this class of veins, for ascertaining
the relation between granite and other mineral
bodies, has been pointed out, § 82 ; and by means
of them it has been shown, that the granite, though
inferior in position, is of more recent formation
than the schistus incumbent on it ; and that the
latter, instead of having been quietly deposited on
the former, has been, long after its deposition and
consolidation, heaved up from its horizontal position, by the liquid body of granite forcibly impelled
against it from below.

* Trans. Royal Society Edin. Vol. III. p. 77, &c.

It has been alleged, in order to take off the force of the argument derived from granite veins, that these veins are formed by infiltration, though, to give any probability to this supposition, it would be necessary to show, that water is able to dissolve the ingredients of granite ; and even if this could be done, the direction which the veins have, in many instances, rising up from the granite, is a proof, as remarked § 82, that they cannot be the effect of infiltration.

Another objection has been thrown out, namely, that the veins here referred to are not of true granite, according to the definition which mineralogists have given of that substance. The force of a fact, however, is not to be lessened by a change of names, or the use of arbitrary definitions. The geneial fact is, that the granitic mass, and the vein proceeding from it, constitute one continuous, and uninterrupted body, without any line of separation between them. The geological argument turns on this circumstance alone; and it is no matter whether the rock be a syenite, a granitelle, or a real granite. The phenomenon speaks the same language, and leads to the same conclusion, whatever be the technical terms the mineralogist employs in describing it.

280. It must, however, be admitted, that a difference of character is often to be observed be-

tween the granite mass and the veins proceeding
from it ; sometimes the substances in the latter are
more highly crystallized than in the former ; sometimes, but more rarely, they are less crystallized,
and, in some instances, an ingredient that enters into the mass seems entirely wanting in the vein.
These varieties, for what we yet know, are not subject to any general rule ; but they have been held
out as a proof, that the masses and the veins are not
of the same formation.   It may be answered, that
a perfect similarity between substances that, on
every hypothesis, must have crystallized in very
different circumstances, is not always to be looked
for ; but the most direct answer is, that this perfect
similarity does sometimes occur, insomuch that, in
certain instances, no difference whatsoever can be
discovered between the mass and the vein, but they
consist of the same ingredients, and have the same
degree of crystallization.   Some instances of this
are just about to be remarked.

281. A strong objection to the supposed origin
of granitic veins from infiltration, and indeed to
their formation in any way but by igneous fusion,
arises from the number of fragments of schistus,
often contained, and completely insulated in those
veins.   How these fragments were introduced into
the fissures of the schistus, and sustained till they
were surrounded by the matter deposited by water,

is very hard to be conceived ; but if they were carried in by the melted granite, nothing is more easily understood.

The following are some of the places where the phenomena of granite veins may be distinctly seen,

282. The island of Arran, remarkable for collecting into a very small compass a great number of the most interesting facts of geology, exhibits many instances of the penetration of schistus by veins of granite. A group of granite mountains occupies the northern extremity of the island, the highest of which, Goatfield, rises nearly to the height of 3000 feet, and on the south side is covered with schistus to the height of 1100. From thence, the line of junction, or that at which the granite emerges from under the schistus, winds, so far as I was able to observe, round the whole group of mountains, with many wavings and irregularities, rising sometimes to a greater, and descending sometimes to a much lower level, than that just mentioned. Along this line, particularly on the south, wherever the rock is laid bare, and cut into by the torrents, innumerable veins of granite are to be seen entering into the schistus, growing narrower as they advance into it ; and being directed, in very many cases, from below upwards, they are precisely of the kind which the infiltration of water could not produce, even were that fluid capable of dissolving the substances which the vein consists

of   From this south face of the mountain, and
from the bed of a torrent that intersects it very
deeply, Dr Hutton brought a block of schistus, of
several hundred weight, curiously penetrated by
granite veins, including in them many insulated
fragments of the schistus.

From this point, the common section of the granite and schistus descends towards the west side of
the mountain, and is visible at the bottom of a deep
glen, (Glen-Rosa,) which detaches Goatfield from
the hills farther to the west.   The junction is laid
bare at several places in the bed of the river which
runs in the bottom of this glen ; and in all of them
exhibits, in a greater or less degree, the appearances of disturbance and violence which have accompanied the injection of the granite veins. Many
circumstances render this spot interesting to a geologist, and, among others, an intersection of the
granite, a little above its junction with the schistus,
by a dike or vein of very compact whinstone.

The same line of junction is found on the opposite, or north-east, side of the mountain, where it
is intersected by another little river, the Sannax,
which on this side determines the base of the mountain.   This junction is no less remarkable than the
other two.

The island of Arran contains, I have no doubt,
many other spots where these phenomena are to be
seen ; but I have had no opportunity of observing

them, nor do I find that Dr Hutton met with any others in his visit to this island.

283. Another series of granite veins is found in Galloway, which was first discovered by Dr Hutton and his friend Mr Clerk, and afterwards more fully explored by Sir James Hall and Mr Douglas, the present Earl of Selkirk. The two last traced the line of separation between a mass of granite and the schistus incumbent upon it, all round a tract of country, about eleven miles by seven, extending from the banks of Loch Ken westward ; and in all this tract they found, " that wherever the junction of the granite with the schistus was visible, veins of the former, from fifty yards, to the tenth of an inch in width, were to be seen running into the latter, and pervading it in all directions, so as to put it beyond all doubt, that the granite of these veins, and consequently of the great body itself, which was observed to form with the veins one uninter-rupted mass, must have flowed in a soft or liquid state into its present position." *   I have only far-ther to add, that some of these veins are remark-able for containing granite, not sensibly different, in any respect, from the mass from which they pro-ceed.

284. In Inverness-shire, between Bernera and Fort Augustus, the same phenomena occur on the

---

* Trans. Royal Society Edin. Vol. III. p. 8.

north side of Loch Chloney, where some granite
mountains rise from under the schistus. In travelling near this place, Lord Webb Seymour and myself were advertised of our approach to a junction
of granite and schistus, by finding among the loose
stones on the road many pieces of schistus, intersected with veins of feldspar and granite. We
walked along this junction for more than a mile ;
and toward the east end, where the road leaves it,
we saw, in the bed of a stream that runs into Loch
Chloney, many beautiful specimens of granitic veins
pervading the schistus, and branching out into very
minute ramifications.

285. The last instance I have to mention from
my own observation, is at St Michael's Mount in
Cornwall. That mount is entirely of granite, thrust
up from under a very hard micaceous schistus, which
surrounds it on all sides. At the base of it, on the
west side, a great number of veins run off from the
granite, and spread themselves like so many roots
fixed in the schistus : they are seen at low water.
In the smaller veins, the granite is of very minute,
though distinct parts ; in the larger, it is more highly crystallized, and is undistinguishable from the
mass of the hill.

Besides the above, Cornwall probably affords
many other instances of the same kind, which I
have not had an opportunity to examine. Such
instances may in particular be looked for at the

Land's End, where a promontory, consisting of a central part of granite, and covered by a micaceous schistus on both sides of it, is cut transversely by the sea coast, and the contact of the granite and schistus of course twice exposed to view.

286. Scotland also affords other examples of granite veins, and some of them have been actually described. Mr Jameson has taken notice of some which he saw in the bottom of the river Spey, at Glen Drummond, in Badenoch, and has represented them in an engraving. * They traverse the strata in various directions, and inclose pieces of the micaceous schistus; and, from the great number of loose blocks which he found, exhibiting portions of such veins, it is probable, that they are very numerous in this quarter. The same mineralogist mentions some instances of similar veins in the Shetland Isles. †

In Ross-shire, Sir George Mackenzie has observed a great variety of granite veins, some of them of large size. One of them, in particular, not far from Coul, when first discovered, was supposed to be a single mass, rising from under the schistus; but, on a more careful examination, has been found to be a part of a great system of veins,

---

* Mineralogy of the Scottish Isles, Vol. II. p. 173.
† Ibid. p. 216.

which intersects the micaceous schistus of this tract in various directions.

287. The granite veins are not the only proof that this stone is more recent than some other productions of the mineral kingdom. Specimens of granite are often found, containing round nodules of other stones, as, for example, of gneiss or micaceous schistus: Such is the specimen of granite containing gneiss, which Werner himself is said to be in possession of, and to consider as a proof, that the schistus is of greater antiquity than the granite. Such also seemed to me some pieces of granite, which I met with in Cornwall, near the Land's End ; and others which I saw in Ayrshire, in loose blocks, on the sea coast between Ayr and Girvan. It is impossible to deny that the containing stone is more modern than the contained. The Neptunists indeed admit this to be true, but allege, that all granite is not of the same formation ; and that, though some granite is recent, the greater part boasts of the highest antiquity which belongs to any thing in the fossil kingdom. This distinction, however, is purely hypothetical ; it is a fiction contrived on purpose to reconcile the fact here mentioned with the general system of aqueous deposition, and has no support from any other phenomenon.

2. Granite of Portsoy.

288. The granite of Portsoy is one of the most singular varieties of this stone, and is remarkable for this circumstance, that the feldspar is the substance which has assumed the figure of its proper crystal, and has given its form to the quartz, so that the latter is impressed both with the acute and obtuse angles belonging to the rhombic figure of the former. The angular pieces of quartz thus moulded on the feldspar, and ranged by means of it in rows, give to this stone the appearance of rude alphabetical writing.

Now, Dr Hutton argued, that substances precipitated from a solution, and crystallizing at liberty, cannot be supposed to impress one another in the manner here exemplified; and that they could do so only when the whole mass acquired solidity at the same time, or at the same time nearly.* Such simultaneous consolidation can be produced in no way that we know of, but by the cooling of a mass that has been in fusion.

289. A granite, brought from Daouria by M. Patrin, and described by him in the Journal de Physique for 1791, p. 295, under the name of *pierre graphique*, seemed to Dr Hutton to have

---

* Theory of the Earth, Vol. I. p. 104.

so great a resemblance to the granite of Portsoy, that he ventured to consider them both as the same stone, and as both containing quartz moulded on feldspar.* It should seem, however, from further explanations, which M. Patrin has since given, that Dr Hutton was mistaken in his conjecture, and that, in the *pierre graphique* of the former mineralogist, the quartz gives its form to the feldspar, preserving in its crystals their natural angle of 120 degrees.† It is impossible, I think, to doubt of the accuracy of this statement; and the graphical stone of Portsoy must therefore be admitted to differ materially from that of Daouria. They are not, however, without some considerable affinity, besides that of their outward appearance; for, though the quartz in the former is generally moulded on the feldspar, the feldspar is also occasionally impressed by the quartz, and sometimes even included in it. They may be considered as varieties of the same species of granite; and the *pierre graphique* of Corsica is probably a third variety, different from them both.

290. It would seem, however, that all these stones lead exactly to the same conclusion. M. Patrin describes his specimen as containing quartz

---

* Trans. Royal Society Edin. Vol. III. p. 83.

† Bibliotheque Britannique, (of Geneva,) 1798, Vol. VIII. Sciences et Arts, p. 78.

crystals, that are for the most part only *cases*, having their interior filled with feldspar. " Le feldspath en masse contient des crystaux quartzeux, qui
n'ont le plus souvent que la carcasse, et dont l'in-
terieur est rempli de feldspath ; souvent il manque
à ces carcasses quelques unes de leurs faces, et souvent la section de cette pierre dans un sens trans-
versal aux crystaux, presente une suite de figures
qui sont des portions d'hexagones, et qui ne resem-
blent pas mal à des caractères Hebraiques." *

These imperfect hexagonal cases of quartz, filled
with feldspar, certainly indicate the crystallization
of substances, which all assumed their solidity at
the same time, and, in doing so, constrained the
figures of one another. To use the words of Dr
Hutton, " whether crystallizing quartz inclose a
body of feldspar, or concreting feldspar determine
the shape of fluid quartz, particularly if we have, as
is here the case, two solid bodies including and included, it amounts to a demonstration, that those
bodies have concreted from a fluid state of fusion,
and have not crystallized, in the manner of salts,
from a solution." †

291. The quartz in granite so generally receives
the impressions of all the other substances, particularly of the feldspar and schorl, and appears to be so

---

* Bibliothèque Britannique, *Ibid.*
† Trans. Royal Society Edin. *ubi supra,* p. 84.

passive a body, that it has been doubted by some mineralogists, whether in this stone it ever assumes its own figure, except where cavities afford room for its crystallization. But it is certain that, beside the Daourian granite just mentioned, there are others, in which the quartz is completely crystallized. Of this sort are some specimens, found in a granite vein on the west side of the hill of St Agnes, in Cornwall. The vein traverses the primitive schistus, of which that hill consists, from south to north nearly: the stone is much decomposed, and the feldspar in general is almost reduced to the state of clay. In this decomposed mass, quartz crystals are found, having the shape of double hexagonal pyramids, perfectly regular and complete. The side of the hexagon, which is the base of the two opposite pyramids, varies from half a tenth to a tenth of an inch in length, and is the same with the altitude of each of the pyramids. In some few specimens, the two pyramids do not rest on the same base, but are separated by a very short, though regular, hexagonal prism. The surfaces of these crystals are rough, and somewhat opaque, with slender spiculæ of schorl frequently traversing them. This roughness is occasioned by slight furrows on the surface of the crystal, very regularly disposed, and parallel to one another, being without doubt impressions from the thin plates of the feldspar, which surrounded the crystal, and

4

slightly indented it. They very much resemble some impressions, remarked by Dr Hutton in the granite of Portsoy, and ascribed by him also to a similar cause. He has represented these in his Theory of the Earth, Vol. I. Plate II. fig. 4. The action and reaction of two crystallizing bodies, hardly admits of a stronger and more unequivocal expression, than in these two instances.

Where the granite was little decomposed, the quartz was not easily disengaged from the mass it was imbedded in, and often broke in pieces before it could be extricated. The crystallization of the quartz, therefore, would not have been discovered, but for the decomposition of the feldspar ; and it is probable, that similar crystallizations exist in many granites where they are not perceived.

292. Some mineralogists are inclined to think, that the regular crystallization of quartz is to be found only in what they call secondary granites, or in those that are of a formation subsequent to the great masses which constitute the granite mountains. It is indeed true, that in the instances given here, both from Cornwall and Daouria, the granites containing quartz crystals are from veins that intersect the primary schistus, and are therefore, on every hypothesis, of a formation subsequent to that schistus. But it does not follow from thence, that they are less ancient than the great masses of unstratified granite ; with these last they are most pro-

bably coeval, nor can there be any reason for thinking the crystallization of quartz a mark of more recent formation than that of feldspar.

3. *Stratification of Granite.*

293. What are the various modes in which granite exists, is a question not absolutely decided
among mineralogists.   1. That it exists as a schistose stone of a fissile texture, in gneiss and *veined
granite*, is on all hands admitted, though in this
state the name of granite is generally withheld from
it.   2. That it exists often without any indication
of a fissile texture, and altogether unstratified, is
likewise acknowledged.   3. That it is found in
veins, intersecting the strata, has been shown above.
The only mode of its existence subject to dispute,
is that in which it is said to be stratified in its outward configuration, but not schistose in its texture.
On this point mineralogists do not perfectly agree :
Dr Hutton did not think that this was a state in
which granite ever appears.   When not schistose
in its structure, he supposed it to be unstratified altogether ; and he considered it as a body which,
like whinstone, was originally in a state of igneous
fusion, and, in that condition, injected among the
strata.   The school of Werner, on the other hand,
maintain, that granite, if not always, is generally

stratified, and disposed in beds, sometimes horizontal, though more frequently vertical, or highly inclined.

In forming an opinion where there are great authorities on opposite sides, a man must trust chiefly to his own observations, and ought to esteem himself fortunate if these lead to any certain conclusion. Mine incline me to differ from Dr Hutton, on the one hand, and from the Neptunists, on the other, as they convince me, that granite does form strata where it has no character of gneiss; and, at the same time, induce me to suspect, that the stratification ascribed by the Neptunists to the granite mountains, is, in many instances, either an illusion, or at least something very different from what, in other stones, is accounted stratification.

294. The first example I ever saw of granite that was stratified, and yet had no character of gneiss, was at Chorley Forest, in Leicestershire. The greater part of that forest has for its base a hornstone schistus, primary and vertical; and, on its eastern border, particularly near Mount Sorrel, are beds of granite, holding the same direction with those of the schistus. The stone is a real granite; it has nothing in its internal structure of a schistose or fissile appearance; and its beds, which it is material to remark, are no thicker than those of the hornstone strata in the neighbourhood. This granite is remarkable, too, for being close to the se-

condary sandstone strata ; I did not see their contact, but traced them within a small distance of one
another ; so that I think it is not likely that any
body of rock intervenes. At the same time that I
state my belief of this rock of granite being in regular strata, I must acknowledge, that a very intelligent mineralogist, who viewed these rocks at the
same time, and whose eye was well practised in geological observation, remained in doubt concerning
them.

295. Another instance of a real granite, disposed in regular beds, but without any character of
gneiss, is one which I saw in Berwickshire, in Lammermuir, near the village of Priestlaw. The little
river of Fassnet cuts the beds across, and renders it
easy to observe their structure. The beds are not
very thick ; they run from about S.S.W. to N.N.E.
like the schistus on either side of them. I was in
company with Sir James Hall when I saw these
rocks ; we examined them with a good deal of attention, and traced them for more than a mile in
the bed of the river ; and, if I mistake not, our opinions concerning them were precisely the same.

296. What exists in two instances may exist in
many, and, after these observations, I should be
guilty of great inconsistency, in refusing to assent·
to the accounts of Pallas, Deluc, Saussure, and
many other mineralogists, who so often represent
granite as formed into strata. In some cases, how-

ever, it is certain, that the stratification they describe is extremely unlike that in the two instances just mentioned, and indeed very unlike any thing that is elsewhere known by the name of stratification. For example, the stratification must be very ambiguous, and very obscurely marked, that was not discovered till after a series of observations, continued for more than twenty years, by a very skilful and distinguishing mineralogist. Yet such undoubtedly is the stratification of Mont Blanc, and of the granite mountains in its neighbourhood, as it escaped the eyes of Saussure, in the repeated visits which he made to them, during a period of no less extent than has just been mentioned. It was not till near the conclusion of those labours, to which the geologists of every age will consider themselves as highly indebted, that, having reached the summit of Mont Blanc, he perceived, or thought that he perceived, the stratification of the granite mountains. The Aiguilles or Needles which border the valley of Chamouni, and even Mont Blanc itself, appeared to be formed of vast tabular masses of granite, in position nearly vertical, and so exactly parallel, that he did not hesitate to call them by the name of strata. Till this moment, these same mountains, viewed from a lower point, had been regarded by him as composed of great plates of rock, nearly vertical indeed, but applied, as it were, round an

axis, and resembling the leaves of an artichoke ; *
and the fissures by which they are separated from
one another, had been considered as effects of waste
and degradation. " But now," (says he, speaking
of the view from the top of Mont Blanc,) " I was
fully convinced, that these mountains are entirely
composed of vast plates of granite, perpendicular to
the horizon, and directed from N.E. to S.W. Three
of these plates, separated from each other, formed
the top of the Aiguille du Midi, and other similar
plates, decreasing gradually in height, compose its
declivity to the south." †

297. Saussure was so strongly impressed with
the appearances of what he accounted regular stratification, such as water only can produce, and such
as must have been in the beginning horizontal, that,
placed as he now was, on one of the highest points
of the earth's surface, he formed the bold concep-
tion, that the summit on which he was standing had
been once buried under the surface, to the depth at
least of half the diameter of the mountain, and horizontally distant from its present place by a line not
less than the whole height of the mountain ; the
granite beds which compose that mountain, having
been raised by some enormous power from their ho-

---

* Voyages aux Alpes, Tom. II. § 910, &c.
† Voyages aux Alpes, Tom. IV. § 1996.

rizontal position, and turned as on an axis, till they were brought into the vertical plane. In this notion, which suits so well with the nature of mountains really composed of vertical strata, and which does credit to the extent of Saussure's views, it is wonderful that he did not see the overthrow of the geological system he had adopted, which is provided with no means whatsoever of explaining these great effects.

Such, then, were the ideas suggested to Saussure, by viewing the mountains of the Alps from the highest of their summits. His great experience, his accurate knowledge of the objects before him, and the power he had acquired of dissipating those illusions, to which, in viewing mountainous tracts, the eye is peculiarly subject, all conspire to give great weight to his opinion. Yet, as this opinion is opposed by that which he himself had so long entertained, before it can be received with perfect confidence, it will require to be verified by new observations. It seems certain, that the beds of rock here described, differ from all ordinary strata, both horizontal and vertical, in the circumstance of their vast thickness, three of them being so large as to form the main body of a mountain. Their parallelism cannot easily be ascertained ; and they have at best but a very slight resemblance to such beds as water is known to produce.

298. Their parallelism is difficult to be ascer-

tained ; for, on account of the magnitude and in-
accessibility of the objects, it is impossible to place
the eye in any situation, where it shall nòt be much
nearer to one part of the planes whereof the parallelism is to be estimated, than to another. Indeed,
one can perceive a cause which may have rendered
the parallelism of the plates of granite which compose the *aiguilles*, more accurate in appearance
than in reality, when viewed from a point so elevated as the summit of Mont Blanc. For, even
on the supposition that the comparison of those
plates to leaves of artichokes was just, and that the
planes of their separation converged toward one
another, in ascending to the top, when they were
viewed from a point more elevated than that top,
this convergency would be diminished, and, by the
force of the perspective, might even be converted
into parallelism. We cannot at present ascertain
what effect this cause of deception may have actually produced.

299. The observations of Saussure concerning
the stratification of granite, are not, however, in
all instances, liable to these objections ; and it
seems to be on much less exceptionable grounds
that he pronounces the granite of St Gothard to
be stratified. The gneiss and micaceous schistus
which constitute the lower part of that mountain,
are succeeded by a granite without any schistose
appearance, but divided into large plates, exactly

parallel to the beds of the former gneiss. These he regards as real strata. On studying them in detail, he says, considerable irregularities were to be observed, but not greater than in the case of limestone or micaceous schistus. * It may be inferred from this, that these plates of granite are not so thick but that they admit of comparison with beds that are known with certainty to be of aqueous formation, and I am therefore disposed to believe, that the granite of St Gothard, in this part at least, is stratified. The transition from gneiss to granite *en masse*, is not uncommon, as Saussure has observed in other instances, and as we are just about to consider more particularly.

300. In the mountains of our own country, some difficulties concerning the stratification of granite have also occurred. In Arran, for instance, the mountain of Goatfield, which I have mentioned above as affording an instance of granite sending out many veins into the schistus, and rivetted, as it were, by means of them to the superincumbent rock, when I visited it, with a view of verifying on the spot the interesting observations which Dr Hutton had there made, appeared to me to be without any vestige of stratification in its granitic part, as did also the whole group of mountains to which it belongs. It was, therefore,

---

* Voyages aux Alpes, Tom. IV. § 1830.

not without a good deal of surprise, that I lately
read, in an account of that island, by a very accurate and ingenious mineralogist, that Goatfield consists of stratified granite. *   The impression which
the appearance of that mountain made on my mind,
is just the reverse ; and though I saw large tabular masses, sometimes nearly vertical, separated by
fissures, they appeared to be much too irregular,
too little extended in length and height, and vastly too much in thickness, to be reckoned the effects
of stratification. For all this, I would by no means
be understood to set my observations in opposition
to those of Mr Jameson.   In my visit to Arran, I
did not direct my inquiries much toward this
point ; the general appearance of the rocks did
not suggest the necessity of doing so, and I was
not perfectly aware how much the stratification of
granite had been insisted on by some mineralogists ;
so that I applied myself entirely to study some
other of the interesting phenomena which this little island offers in so great abundance.   I therefore carry my confidence in the appearances which
seemed to indicate a want of stratification in the
granite of Arran no further than to remain scepti-
cal both as to Mr Jameson's conclusions and my
own, till an opportunity shall occur of verifying
the one or the other by actual observation.

---

* Mineralogy of the Scottish Isles, Vol. I. p. 35, 36.

301. The stratification of granite, though it made no part of Dr Hutton's system, does by no means embarrass his theory with any new difficulty. Rocks, of which the parts are highly crystallized, are already admitted as belonging to the strata, and are exemplified in marble, gneiss, and veined granite. In the two last, we have not only stratification, but a schistose, united with a crystallized structure, and the effects of deposition by water, and of fluidity by fire, are certainly no where more singularly combined. The stratification of these substances is therefore more extraordinary than even that of the most highly crystallized granite. Neither the one nor the other can be explained but by supposing, that while such a degree of fluidity was produced by heat, as enabled the body when it cooled to crystallize, the whole mass was kept in its place by great pressure acting on all sides, so that the shape was preserved as originally given to it by the sea. As we cannot, however, suppose, that the intensity of the heat, or the fusibility of the substance through all the parts of a stratum, were precisely the same, we may expect to find in the same stratum, or in the same body of strata, that in some parts the marks of stratification are completely obliterated, while in others they remain entire. It is thus that *veined granite*, or what I think should be called granitic schistus, often graduates into granite *in mass*, that is, gra-

nite without any schistose or fissile texture. Saussure says, that to be veined or not veined, is an af-
fection of granite, that seems, in many cases, accidental ; * as, in the midst of rocks of that substance, most clearly fissile, large portions appear
without any vestige of stratification.   Of this phenomenon, which is frequent in the Alps, instances
are also to be met with in the granite rocks of
Scotland, and the adjacent isles ; and I know that
Dr Hope, in a mineralogical excursion which he
lately made among the Hebrides, observed many
interesting and curious examples of it.   Indeed,
when rocks were so much fused as to crystallize,
and so compressed, at the same time, as to remain
stratified, they were evidently on the verge of
change ; two opposite forces were very nearly balanced, and each carried as far as it could go without entirely overcoming the other ; so that a small
alteration in the conditions may have made a great
alteration in the effects.   Hence a sudden transition from a stratified to an unstratified texture,
which is only found in rocks highly crystallized,
and such as have endured the most violent action
of the mineralizing powers.

302. Now, though the stratification of granite,
or the mixture of the stratified with the unstratified rocks of that genus, is not only reconcileable

* Voyages aux Alpes, Tom. IV. § :143.

with the principles of the Huttonian geology, but
might even have been deduced as a corollary from
those principles, before it was actually observed, it
may be considered as inconsistent with the theory
of granitic veins that has just been given.   A stratum, though soft or fluid, could not invade the surrounding strata with violence, nor send out veins
to penetrate into them.   It might, if strongly
compressed by another stratum less fluid than itself, fill up any fissures or cracks that were in that
other, but this would hardly produce such large
veins, and of such considerable length, as often
penetrate from the granite into the schistus, nor
could it give rise to any appearance of disturbance.
If, therefore, veins were found proceeding from
such stratified granite as that of Chorley Forest or
Lammermuir, I should think, that the explanation of them was still a *desideratum* in geology.
The Neptunian theory of infiltration would indeed
be as applicable to them as to any other veins ; for
it is but little affected by the condition of the phenomena to be explained.   Indeed, it is very difficult to set any limits to the explanations which this
theory affords ; and it would certainly puzzle a
Neptunist, to assign any good reason why infiltration has not produced veins of one schistus running
into another, or veins of schistus running into granite, as well as of granite running into schistus.

He will find it a hard task to restrain the activity of his theory, and to confine its explanations to those things that really exist.

303. As the Huttonian system cannot boast of theories of equal versatility, it would be not a little embarrassed to account for veins of great magnitude proceeding from a rock distinctly stratified, and accompanied with marks of having disturbed the rocks through which they pass. I am, however, inclined to believe, that this embarrassment will never occur; and that the granite veins do not proceed from the rocks that are really stratified, but from such as have never been deposited by water, and where the appearances of stratification, if there are any, are altogether illusory. This anticipation, however, requires to be verified by future observation ; and it remains to be seen, whether granitic veins ever accompany real granitic strata, or are peculiar to those in which the appearances of regular beds are either ambiguous, or are entirely wanting. The decision of this question is an object highly worthy of the attention of geologists.

304. An argument, directed at once against the igneous origin and unstratified nature of all granite, is given in a work already mentioned. " If granite had flowed from below, how does it happen, that, after it had burst through the strata of mica-

11

ceous schistus, &c. it did not overflow the neighbouring country? If this hypothesis were true,
Mont Blanc could never have existed." *

A theory is never more unfairly dealt with, than
when those parts are separated which were meant
to support one another, and each left to stand or
fall by itself. This, however, is precisely what is
done in the present instance; for Dr Hutton's
theory of granite would not deserve a moment's
consideration, if it were so inartificially constructed, as to suppose that granite was originally fluid,
and yet to point out no means of hindering this
fluid from diffusing itself over the strata, and settling in a horizontal plane. The truth is, that his
theory, at the same time that it conceives this stone
to have been in fusion, supposes it to have been,
in that state, injected among the strata already consolidated; to have heaved them up, and to have
been formed in the concavity so produced, as in a
mould. Thus Mont Blanc, supposing that it is unstratified, is understood to consist of a mass that was
melted by subterraneous heat under the strata, and
being impelled upwards by a force, that may stand
in some comparison with that which projected the
planets in their orbits, heaved up the strata by
which it was covered, and in which it remained included on all sides.

---

* Mineralogy of the Scottish Isles, Vol. II. p. 166.

305. The covering of strata, thus raised up, may
have been burst asunder at the summit, where the
curvature and elevation were the greatest; but the
melted mass underneath may have already acquired
solidity, or may have been sustained by the beds
of schistus incumbent on its sides. This schistus,
forming the exterior crust, was immediately acted
on by the causes of waste and decomposition, which
have long since stripped the granite of a great part
of its covering, and are now exercising their power
on the central mass. That even Mont Blanc itself, as well as other unstratified mountains, was
once covered with schistus, will appear to have in
it nothing incongruous, when we consider the height
to which the schistus still rises on its sides, or in
the adjacent mountains; and when we reflect, that,
from the appearances of waste and degradation
which these mountains exhibit, it is certain, that
the schistus must have reached much higher than
it does at present.

It is obvious, therefore, that when the corresponding parts are brought together, and placed in
their natural order, no room is left for the reproach, that this system is inconsistent with the
existence of granite mountains. I have no pleasure in controversial writing; and, notwithstanding the advantages which a weak attack always
gives to a defender, I cannot but regret, that Dr

Hutton's adversaries have been so much more eager to refute than to understand his theory.

---

306. A remark which Dr Hutton has made on the quantity of granite that appears at the surface, compared with that of other mineral bodies, has been warmly contested. Having affirmed, that the greater part of rocks bear marks of being form-ed from the waste and decomposition of other rocks, he alleges that granite, (a stone which does not contain such marks) does not, for as much as appears from actual observation, make up a tenth, nor perhaps even a hundredth part of the mineral kingdom. * Mr Kirwan contends, that this is a very erroneous estimate, and that the quantity of granite visible on the surface, far exceeds what is here supposed. † The question is certainly of no material importance to the establishment of Dr Hutton's theory : it is evident, too, that an esti-mation, which varies so much as from a tenth to a hundredth part, cannot have been meant as any thing precise ; yet it may not be quite superfluous to show, that the truth probably lies nearer to the least than the greatest of the limits just mentioned.

---

* Theory of the Earth, Vol. I. p. 211.
† Geol. Essays, p. 480.

307. Though granite forms a part, generally the central part, of all the great chains of mountains, it usually occupies a much less extent of surface than the primary schistus. Thus in the Alps, if a line be drawn from Geneva to Ivrea, it will be about eighty-five geographical miles in length, and will measure the breadth of this formidable chain of mountains, at the place of its greatest elevation. Now, from the observations of Saussure, who cross-ed the Alps exactly in this direction, it may be collected, that less than nine miles of this line, or not above a tenth part of it, in the immediate vici-nity of Mont Blanc, is occupied by granite.

308. In some sections of the Alps, no granite at all appears. Thus, in the route from Chambery to Turin, across Mont Cenis, which measures by the road not less than ninety miles, no granite is found, at least of that kind which is distinctly in mass, and different from gneiss or veined granite. *

309. In some other places of the same moun-tains, the granite is more abundant. A line from the lake of Thun, along the course of the Aar, and over the mountains to the upper end of Lago Mag-giore, crosses a very elevated tract, and passes by the sources of the Rhone, the Rhine, and the Ticino, which last runs into the Po. A good deal of granite is discovered here, in the mountains of

* Voyages aux Alpes, Tom. III. § 1190, &c.

Grimsel and St Gothard ; but by far the greater part of it is the veined granite, the granite in mass being confined chiefly to the north side of the Grimsel.   Both together do not occupy more than one-third of the line, and therefore the latter less than one-sixth.

310. The essay on the mineralogy of the Pyrenees, by the Abbé Palasso, contains a mineralogical chart of those mountains.   From this chart I have found, by computation, that the granite does not occupy one-fifth of the horizontal surface on the north side of the ridge, reckoning from one end of it to the other.   Indeed, many great tracts, even of the central parts of the Pyrenees, contain no granite whatsoever; and not a few of the highest mountains consist entirely of calcareous schistus. A large deduction should be made from the fraction $\frac{1}{5}$, on account of the substances unknown, which, from the construction of the chart, are often confounded with the granitic tract.

311. I might add other estimations of the same kind, all confessedly rude and imperfect, but still conveying, by means of numbers, a better idea of the limit to which our knowledge approximates, than could be done simply by words ; and, on the whole, it would appear, that if we state the proportion of granite to schistus to be that of one to four, we shall certainly do no injustice to the extent of the former.

It remains to form a rough estimate from maps, and from the accounts of travellers, of what proportion of the earth's surface consists of primary, and what of secondary rocks. After supplying the want of accurate measurement by what appeared to me the most probable suppositions, I have found, that about $\frac{1}{18}$ of the surface of the old continent may be conceived to be occupied by primitive mountains; of which, if we take one-fifth, we have $\frac{1}{90}$ for the part of the surface occupied by granite rocks, which differs not greatly from the least of the two limits assigned by Dr Hutton.

312. In estimating the granite of Scotland, Dr Hutton has certainly erred considerably in defect, * and Mr Kirwan, who always differs from him, is here nearest the truth; though he is right purely by accident, as the information on which he proceeds is vague and erroneous.

The places in Scotland where granite is found, are very well known; but the extent of some of the most considerable of them is not accurately ascertained. In the southern parts, except the gra-

---

* Dr Hutton in this case no doubt made a very loose estimate. He says, the granite does not perhaps occupy more than a 500dth part of the whole surface. The whole surface of Scotland is not much more than 23,000 geographical miles, the 500dth part of which is exactly 46; and this is exceeded by the granite in Kirkcudbrightshire alone, as may be gathered from what is said § 283.

nite of Galloway, which is found in two pretty large insulated tracts, there is no other of any magnitude. The granite of the north extends over a large district. If we suppose a line to be drawn, from a few miles south of Aberdeen to a few miles south of Fort William, it will mark out the central chain of the Grampians in its full extent, passing over the most elevated ground, and by the heads of the largest rivers, in Scotland. Along this line there are many granite mountains, and large tracts in which granite is the prevailing rock. There are, however, large spaces also in which no granite appears, though, if we were permitted to speak theoretically, and if the question did not entirely relate to a matter of observation, we might suppose, that, in no part of this central ridge is the granite far from the surface, notwithstanding that in some places it may be covered by the schistus.

313. A great part of the Grampian mountains is on the south side of the line just mentioned, but hardly any granite is found in this division of them, except such veins as those of Glentilt. On the north side of the line, the granite extends in various directions; and, if from Fort William a line is drawn to Inverness, the quadrilateral figure, bounded on two sides by these lines, and on the other two by the sea, will be found to contain much granite, and many districts consisting entirely of that stone.

This is in fact the great granite country of Scot-

land : it is a large tract, containing about 3170
square geographical miles, or about a seventh part of
the whole : but the proportion of it occupied by
granite cannot at present be ascertained with any
exactness, nor will, till some mineralogist shall find
leisure to examine the courses of the great rivers,
the Dee, the Spey, &c. which traverse this country.
If we call it one-fourth of the whole surface, its extent is certainly not underrated, and will amount to
790 square miles nearly ; to which adding 150, as
a very full allowance for all the other granite contained in Scotland, exclusive of the isles, we shall
have 940 square miles, between a twenty-fourth
and twenty-fifth part of the surface of the whole.

This computation, it must be observed, aims at
nothing precise, but I think it is such, that a more
accurate survey would rather diminish than increase the proportion assigned in it to the granite
rock.

314. This result may perhaps fall as much short
of Mr Kirwan's notion, as it exceeds the estimate
made by Dr Hutton.   If it shall not, and if the
former has, in this instance, come nearest the truth,
it cannot be ascribed to the accuracy of his information, or the soundness of the principles which directed his research. Mr Williams, whom he quotes,
was a miner, of great skill and experience in some
branches of his profession, to which, if he had confined himself, he might have written a book full of

useful information.   What he says on the subject
of granite, is, in the main I believe just ; but it is
far too general to authorize the conclusion which
Mr Kirwan derives from it.   Dr Ash, for whose
judgment I have great respect, cannot, I think,
have meant, when he used the expression granitic
rocks, to describe granite strictly so called.   He
says, in the passage quoted by Mr Kirwan, that
" from Galloway, Dumfries, and Berwick, there is
a chain of mountains, commonly schistose, but
often also granitic."   Now, the fact is, that the
great belt of primary rock, here alluded to, which
traverses the south of Scotland, consists of vertical
schistus of various kinds ; but except in Galloway,
and again in Lammermuir, near Priestlaw, it appears, as already mentioned, to contain no granite
whatsoever.   If the German mineralogist quoted
by Mr Kirwan, when he says that the Grampian
mountains consist of micaceous limestone, gneiss,
porphyry, argillite, and granite, alternating with
one another, means only to affirm that all these
stones are found in the Grampians, he is certainly
in the right, and the catalogue might easily be enlarged ; but, if he either means to say, that these
are nearly in equal abundance, or that the granite
is commonly found in strata alternating with other
strata, I must say, that these are propositions quite
contrary to any thing I have ever seen or heard of
those mountains.   But it is probable that this is

not meant, and that the fault lies in understanding
the expressions much too literally. Mr Kirwan
accuses Dr Hutton of not knowing where to look
for the granite ; not aware of how much, notwithstanding any error committed in the present estimate, he was skilled in the art of mineralogical observation ; an art, which those who have not practised do not always know how to appreciate.   But,
however imperfect Mr Kirwan's knowledge of this
subject has been, he has here had the good fortune
to correct a mineralogist of very superior information.   The mere disposition to oppose is not always without its use : no man is in every thing free
from error, and, to controvert indiscriminately all
the opinions of any individual, is an infallible secret
for being sometimes in the right.

NOTE XVI. § 100.

Rivers and Lakes.

315. Rivers are the causes of waste most visible
to us, and most obviously capable of producing
great effects.   It is not, however, in the greatest
rivers, that the power to change and wear the surface of the land is most clearly seen.   It is at the
heads of rivers, and in the feeders of the larger
streams, where they descend over the most rapid

slope, and are most subject to irregular or temporary increase and diminution, that the causes which tend to preserve, and those that tend to change the form of the earth's surface, are farthest from balancing one another, and where, after every season, almost after every flood, we perceive some change produced, for which no compensation can be made, and something removed which is never to be replaced. When we trace up rivers and their branches toward their source, we come at last to rivulets, that run only in time of rain, and that are dry at other seasons. It is there, says Dr Hutton, that I would wish to carry my reader, that he may be convinced, by his own observation, of this great fact, *that the rivers have, in general, hollowed out their valleys.* The changes of the valley of the main river are but slow; the plain indeed is wasted in one place, but is repaired in another, and we do not perceive the place from whence the repairing matter has proceeded. That which the spectator sees here, does not therefore immediately suggest to him what has been the state of things before the valley was hollowed out. But it is otherwise in the valley of the rivulet; no person can examine it without seeing, that the rivulet carries away matter which cannot be repaired, except by wearing away some part of the surface of the place upon which the rain that forms the stream is gathered. The remains of a former state are here visible; and

we can, without any long chain of reasoning, compare what has been with what is at the present moment.   It requires but little study to replace the
parts removed, and to see nature at work, resolving
the most hard and solid masses, by the continued
influences of the sun and atmosphere.*   We see
the beginning of that long journey, by which heavy
bodies travel from the summit of the land to the
bottom of the ocean, and we remain convinced,
that, *on our continents, there is no spot on which
a river may not formerly have run.* †

316. The view thus afforded of the operations,
in their nascent state, which have shaped out and
fashioned the present surface of the land, is necessary to prepare us for following them to the utmost
extent of their effects.   From these effects, the
truth of the proposition, that rivers have cut and
formed, not the beds only, but the whole of the valleys, or rather system of valleys, through which they
flow, is demonstrated on a principle which has a
close affinity to that on which chances are usually
calculated, § 99.   In order to conceive rightly the
course of a great river, and the communication subsisting between the main trunk and its remotest
branches, let us take the instance of the Danube,
and cast our eyes on one of the maps constructed

---

* Theory of the Earth, Vol. II. p. 294.
† Ibid. p. 296.

by Marsigli, for illustrating the natural history of
that great river.* When it is considered, that
over all the vast and uneven surface, which reaches
from the Alps to the Euxine, and from the mountains of Crapack to those of Hæmus, a regular
communication is kept up between every point
and the line of greatest depression, in which the
river flows, no one can hesitate to acknowledge,
that it is the agency of the waters alone which has
opened them a free passage through all the intrica-
cies of this amazing labyrinth. In effect, suppose
this communication to be interrupted, and that
some sudden operation of nature were to erect a
barrier of mountains to oppose the Theise or the
Drave, as they rolled their waters to the Danube.
From this what could possibly result, but the dam-
ming up of those rivers till their waters were deep,
or high enough to find a vent, either under the
bases or over the tops of the opposing ridge. Thus
there would be formed immense lakes and immense
cataracts, which, by filling up what was too low,
and cutting down what was too high, would in time
restore such a uniform declivity of surface as had
before prevailed. Just so in the times that are
past, whatever may have been the irregularities of
the surface at its first emerging from the sea, or
whatever irregularities may have been produced in

* Histoire du Danube, Tom. I. Tab. 34.

it by subsequent convulsions, the slow action of the
streams would not fail in time to create or renew
a system of valleys communicating with one another, like that which we at present behold.   Water, in all circumstances, would find its way to the
lowest point ; though, where the surface was quite
irregular, it would not do so till after being dammed
up in a thousand lakes, or dashed in cataracts over
a thousand precipices.   Where neither of these is
the case ; and where the lake and the cataract are
comparatively rare phenomena ; there we perceive
that constitution of a surface, which water alone, of
all physical agents, has a tendency to produce ; and
we must conclude, that the probability of such a
constitution having arisen from another cause, is,
to the probability of its having arisen from the running of water, in such a proportion as unity bears
to a number infinitely great.

317. The courses of many rivers retain marks
that they once consisted of a series of lakes, which
have been converted into dry ground, by the twofold operation of filling up the bottoms, and deepening the outlets.   This happens, especially, when
successive terraces of gravelly and flat land are
found on the banks of a river, § 100.   Such platforms, or *haughs* as they are called in this country,
are always proofs of the waste and *detritus* produced
by the river, and of the different levels on which it
has run ; but they sometimes lead us farther, and

make it certain, that the great mass of gravel which forms the successive terraces on each side of the river, was deposited in the basin of a lake. If, from the level of the highest terrace, down to the present bed of the river, all is alluvial, and formed of sand and gravel, it is then evident, that the space as low as the river now runs must have been once occupied by water; at the same time, it is clear, that water must have stood, or flowed as high at least, as the uppermost surface of the meadow. It is impossible to reconcile these two facts, which are both undeniable, but by supposing a lake, or body of stagnant water, to have here occupied a great hollow, (which by us must be held as one of the original inequalities of the globe, because we can trace it no farther back,) and that this hollow, in the course of ages, has been filled up by the gravel and alluvial earth brought down by the river, which is now cutting its channel through materials of its own depositing. There is no great river that does not afford instances of this, both in the hilly part of its course, and where it descends first from thence into the plain. Were there room here for the minuter details of topographical description, this might be illustrated by innumerable examples.

318. It is said above, that the water must have run or stood, in former times, as low as the present bottom of the river; but there is often clear evidence, that it has run or stood much lower, because

the alluvial land reaches far below the present level
of the river.   This is known to hold in very many
instances, where it has happened that pits have
been sunk to considerable depths on the banks of
large rivers.   By that means, the depth of the alluvial ground, under the present bed of the river,
has been discovered to be great ; and from this
arises the difficulty, so generally experienced, of
finding good foundations for bridges that are built
over rivers in large valleys, or open plains, the
ground being composed of travelled materials to an
unknown depth, without any thing like the native
or solid strata.   In such cases, it is evident, that
formerly the water must have been much lower, as
well as much higher, than its present level, and
this is only consistent with the notion, that the
place was once occupied by a deep lake.

319. If, following the light derived from these
indications, we go back to the time when the river
ran above the highest of those levels at which it has
left any traces of its operations, we shall see it composed of a series of lakes and cataracts, from which,
by the filling up of the one, and the wearing down
of the other, the waters have at length worked out
to themselves a quiet and uninterrupted passage to
the ocean.   We may, indeed, on good evidence, go
back still farther than the succession of such meadows or terraces, as are above mentioned, will carry us, and may consider the whole valley, or *trough*

of the river, as produced by its own operations. The original inequalities of the surface, and the disposition of the strata, must no doubt have determined the water courses at first; but this does not hinder us from considering the rivers as having modified and changed those inequalities, and as the *proximate* causes of the shape and configuration which the surface has now assumed.

320. From this gradual change of lakes into rivers, it follows, that a lake is but a temporary and accidental condition of a river, which is every day approaching to its termination; and the truth of this is attested, not only by the lakes that have existed, but also by those that continue to exist. Where any considerable stream enters a lake, a flat meadow is usually observed increasing from year to year. The soil of this meadow is disposed in horizontal strata: the meadow is terminated by a marsh; which marsh is acquiring solidity, and is soon to be converted into a meadow, as the meadow will be into an arable field. All this while the sediment of the river makes its way slowly into the lake, forming a mound or bank under the surface of the water, with a pretty rapid slope toward the lake. This mound increases by the addition of new earth, sand, and gravel, poured in over the slope; and thus the progress of filling up continually advances.

321. In small lakes, this progress may easily be

traced ; and will be found singularly conspicuous in
that beautiful assemblage of lakes, which so highly
adorns the mountain scenery of Westmoreland and
Cumberland. Among these a great number of instances appear, in which lakes are either partially
filled up, or have entirely disappeared. In the Lake
of Keswick, we not only discover the marks of filling up at the upper end, which extend far into Borrowdale, from which valley a small river flows into
the lake ; but we have the clearest proof, that this
lake was once united to that of Bassenthwaite,
and occupied the whole valley from Borrowdale to
Ouse-Bridge. These two lakes are at present joined only by a stream, which runs from the former into the latter, and their continuity is interrupted by a considerable piece of alluvial land,
composed of beds of earth and gravel, without rock,
or any appearance of the native strata. This separation, therefore, seems no other than a *bar*, formed by the influx of two rivers, that enter the valley
here from opposite sides, the Greeta from the east,
and Newland's water from the west. The surface
of this meadow is at present twelve or fifteen feet
at least above the level of either lake ; and a quantity of water of that depth must therefore have been
drawn off by the deepening of the issue at Ouse-
Bridge, through which the water of both lakes
passes, in its way to the ocean.

Many more examples, similar to this, may be

collected from the same lakes; there are indeed few places from which, in this branch of geology, more information may be collected.

322. The larger lakes exemplify the same progress. Where the Rhone enters the Lake of Geneva, the beach has been observed to receive an annual increase; and the Portus Valesiæ, now Port Valais, which is at present half a league from the lake, was formerly close upon its bank. Indeed, the sediments of the Rhone appear clearly to have formed the valley through which it runs, to a distance of about three leagues at least from the place where the river now discharges itself into the lake. The ground there is perfectly horizontal, composed of sand and mud, little raised above the level of the river, and full of marshes. The deposition made by the Rhone after it enters the lake, is visible to the eye; and may be seen falling down in clouds to the bottom.

The great lakes of North America are undergoing the same changes, and, it would seem, even with more rapidity. As the rivers, however, which supply these vast reservoirs, are none of them very great, the filling up is much less remarkable than the draining off of the water, by the deepening of the outlet. An intelligent traveller has remarked, that in Lake Superior itself the diminution of the waters is apparent, and that marks can be discovered on the rocks, of the surface hav-

ing been six feet higher than it is at present. In
the smaller lakes this diminution is still more evident. * In some of those far inland, the ground all
round appeared to the same traveller to be the deposit from the rivers, of which the lakes themselves
may be considered as a mere expansion. †

823. In order to give uniform declivities to the
rivers, the lakes must not only be filled up or drained, but the cataract, wherever there is one, must
be worn away. The latter is an operation in all
cases visible. The stream, as it precipitates itself
over the rocks, hurries along with it, not only sand
and gravel, but occasionally large stones, which
grind and wear down the rock with a force proportioned to their magnitude and acceleration. The
smooth surface of the rocks in all waterfalls, their
rounded surface, and curious excavations, are the
most satisfactory proofs of the constant attrition
which they endure; and, where the rocks are
deeply intersected, these marks often reach to a
great height above the level on which the water
now flows. The phenomena, in such instances,
are among the arguments best calculated to remove all incredulity respecting the waste which ri-

---

* Mackenzie's Voyages through the Continent of North
America to the Frozen and Pacific Oceans, p. xlii. and
xxxvi.

† Ibid. p. 122.

vers have produced, and are continuing to produce. They suffer no doubt to remain, that the
height and asperity of every waterfall are continually diminishing ; that innumerable cataracts are
entirely obliterated ; that those which remain are
verging toward the same end, and that the Falls
of Montmorenci and Niagara must ultimately disappear.

324. Though there can be no doubt of the justness of the preceding conclusions, when applied to
lakes in general, some apparent exceptions occur,
in which the progress of draining and filling up
seems to have been suspended, or even to have
gone in a contrary direction. These exceptions
consist of the lakes which appear to have received
a greater quantity of materials than was sufficient
to have filled them up. Such, for example, is the
Lake of Geneva, which receives the Rhone de
scending from the Valais, one of the deepest and
longest valleys on the surface of the earth. Now,
if this valley, or even a large proportion of it, had
been excavated by the Rhone itself, as our theory
leads us to suppose, the lake ought to have been
entirely filled up, because the materials brought
down by the river seem to be much greater than
the lake, on any reasonable supposition concerning its original magnitude, can possibly have received. What, then, it may be said, has become
of all that the Rhone has brought down and de-

posited in it ? The lake, at this moment, retains, in some places, the depth of more than 1000 feet; and yet, of all that the Rhone carries into it, nothing but the pure water issues. If it has been continuing to diminish, both in superficial extent and in depth, from the time when the Rhone began to run into it, what must have been its original dimensions ?

I cannot pretend to remove entirely the difficulty which is here stated ; yet I think the following remarks may go some length in doing so.

325. It is certain, that from the present state of the Lake of Geneva, and of the ground round it, we can hardly draw any inference as to its original dimensions. Saussure has traced, with his usual skill, the marks of the course of the Rhone, on a level greatly above the present ; and, by observations on the side of Mont Saleve, has found proofs of the running of water, at least 200 toises above the present superficies of the lake. But, if ever the superficies of the lake stood at this height, or at this height nearly, though we can conjecture but little concerning the state of the adjacent country, which no doubt was also on a higher level, the lake may very well be supposed to have been of far greater dimensions than it is now. It may have occupied the whole space from Jura to Saleve, and included the Lake of Neufchatel ; so that it may have been of magnitude sufficient to receive the

spoils of the Valais, which, as the surface of its waters lowered, may have been washed away and carried down to the sea. Thus it may have afforded a temporary receptacle for the *débris* of the Alps, and may have served for an *entrepôt*, as it were, where those debris were deposited, before they were carried to the place of their ultimate destination.

326. But the great depth which the lake has at present, still remains to be explained, because no mud or gravel could be carried beyond the gulf, of a thousand feet deep, which was here ready to receive it. The reality of this difficulty must be acknowledged; and some cause seems to act, if not in the generation, yet certainly in the preservation of lakes, with which we are but little acquainted. We can indeed imagine some causes of that kind to occur in the course of the degradation of the land, which may produce new lakes, or increase the dimensions of the old. The wearing away of a stratum, or body of strata, may lay bare, and render accessible to the water, some beds of mineral substances soluble in that fluid. The district, for instance, in Cheshire, which contains rock salt, extends over a tract of fourteen or fifteen miles, and is covered by a thick stratum of clay, more or less indurated, which defends the salt from the water at the surface, and preserves the whole mass in a state of dryness. Should this covering be broke open

by any natural convulsion, or should it be worn away, as it must be in the progress of the general detritus, the water would gain admission to the saline strata, would gradually dissolve them, and form of course a very deep and extensive lake, where all was before dry land. This event is not only possible, but it should seem, that in the course of things it must necessarily happen.

327. Something of this kind may have taken place in the track of the Rhone, and may have produced the Leman Lake. It is not impossible, that, at a very remote period, the Rhone descended from the Alps without forming any lake, or at least any lake of which the remains are now existing ; and this supposition, which is more probable than that of § 325, we shall soon find to be conformable to appearances of another kind. The river may have wore away the secondary limestone strata over which it took its course after it left the schistus of the mountains ; and, in doing so, may have reached some stratum of a saline nature, and this being washed out, may have left behind it a lake, which is but modern compared with many of the revolutions that have happened on the surface of the earth. *

---

* There are salt springs at Bex, near Aigle, about ten miles from the head of the lake : saline strata, therefore, are probably at no great distance.

8

This explanation is no doubt hypothetical ; but it is proposed in one of those cases, in which hypothetical reasonings are warranted by the strictest rules of philosophical investigation.   It is proposed in a case, where the causes visible to man seem inadequate to the effect, and where we must therefore have recourse to an agent that is invisible.   If the operations ascribed to this agent are conformable to the analogy of nature, it is all that can in reason be required.

328. Another circumstance may also influence the generation and preservation of lakes ; but it is also one with which we are but little acquainted. The strata, and indeed the whole body of mineral substances which forms the basis of our land, have been raised up from the bottom of the sea, by a progress that should seem in general to have been gradual and slow.   Appearances, however, are not wanting, which show, that this progress is not uniform ; and that both rising and sinking in the surface of the land, or in the rocks which are the base of it, have happened within a period of time, which is by no means of great extent.   In this progress, the elevations and depressions may not be the same for every spot.   They may be partial, and one part of a stratum, or body of strata, may rise to a greater height, or be more depressed, than another.   It is not impossible, that this process may

affect the depth of lakes, and change the relative level of their sides and bottom.

329. All lakes, however, do not involve the difficulty which the preceding conjectures are intended to remove. The great lakes of North America do not, for instance, receive their supply from very large rivers. Of course, it is not from a tract great in comparison of themselves, that the waste and detritus is brought down into them; and it seems not at all wonderful, that, without being filled up, they have been able to receive it. The same, in a degree at least, is true of many other lakes,

It should also be considered, that we may err greatly in the estimate we make of the materials actually carried down and deposited in any lake. To judge of their entire amount, we should know the original form of the inequalities on the earth's surface; of the quantity of depression which existed, independently of the rivers; and though, in general, these original inequalities may be overlooked, and the present considered as made by the running of water, yet, in particular instances, this may be far from true. The Valais, for example, which we consider as the work of the Rhone, may, when the Alps rose out of the sea, have included many depressions of the surface, which the river joined together, and, from being a series of lakes, formed into one great valley.

330. The mouths by which rivers on bold rocky coasts discharge their waters into the sea, afford a very striking confirmation of the conclusions concerning the general system of waste and degradation which have been drawn above. At these mouths we usually see, not only the bed of the river, but frequently a considerable valley, cut out of the solid rock, while that rock preserves its elevation, and its precipitous aspect, wherever it is not intersected by a run of water. No convulsion that can have torn asunder the rocks ; no breach that can have been made in them, antecedent to the running of the waters, will account for the circumstance of every river finding a corresponding opening, by which it makes its way to the sea ; for that opening being so nearly proportional to the magnitude of the river, and for such breaches never occurring but where streams of water are found.

331. The actual survey of any bold and rocky coast, will make this clearer than any general statement can possibly do. Let us take, for an example, the coast of the British Channel, from Torbay to the Land's End, which is faced by a continued rampart of high cliffs, formed of much indurated and primeval rock. If we consider the breaches in this rampart, at the mouths of the Dart, of the Plym and Tamer, of the river at Fowey, of the Fal, the Hel, &c. it will appear perfectly clear,

that they have been produced by their respective
streams.   Where there is no stream, there is no
breach in the rock, no softening in the bold and
stern aspect which this shore every where presents
to the ocean.   If we look at the smaller streams,
we find them working their way through the cliffs
at the present moment; and we see the steps by
which the larger valleys of the Dart and the Tamer
have been cut down to the level of the sea.   If we
would have still clearer evidence, that no breaches
made antecedently to the running of the rivers
have opened a way for them, we need only look to
the opposite side, or northern shore, of the same
promontory, where we also find a series of outlets,
all originating in the ridge of the country, and becoming deeper as they approach the sea, but altogether unconnected with the openings on the south
side; and this could hardly have been the case,
had they been the effects of previous concussions, or
of any peculiarity in the original structure of the
rocks.

332. In contemplating such coasts as these, when
we go back to the time when the rivers ran upon a
level as high as the highest of the cliffs on the sea
shore, we must suppose, that the land then extended many miles farther into what is now occupied by
the sea.   When at Plymouth, for instance, the
Tamer and the Plym flowed on the level of Mount
Edgecombe or of Staten Heights, if the rivers ran

with a moderate declivity into the sea, the coast must have advanced many miles beyond its present line. Thus the land, when higher, was also more extended, and the limits of our island in that ancient state, were doubtless very different from these by which it is at present circumscribed.

If with the same views we consider any other of the bold coasts which the map of the world presents us with, we shall quickly remark, that wherever a deep intersection of the sea is made into the land, as on the western shores of our own island, or on those of Norway, a river runs in at the head of it, and points out by what means such inlets are formed, viz. by the united powers of the sea and of the land, the waters of the latter having opened the way by which those of the former have penetrated so far into the country.

333. It is not meant assuredly to deny the irregularities of the sea coast, as it may have originally existed ; these irregularities no doubt determined the initial operations of that waste and decay, by which, in process of time, they were themselves entirely effaced. The line of our coasts may be compared to one of those curves, which are sometimes treated of in the higher geometry, where the ordinates are functions, not only of their abscissæ, but also of the time elapsed since a certain epocha. The form of the curve at that epocha, or when the time began to flow, corresponds to the original

form of the sea coast, on its emerging from the
ocean, and before the powers of wasting and decay
had begun to act upon it.   To speak strictly, the
original figure, in both cases, influences all the subsequent ; but the farther removed from it in point
of time, the less is that influence ; so that, in physical questions, and for the purpose of such approximations as suit the imperfection of our knowledge, the consideration of the original figure may
be wholly left out.

NOTE XVII. § 105.

Remains of Decomposed Rocks.

334. The plain of Crau was the Campus Lapi-
deus of the ancients ; and, as mythology always
seeks to connect itself with the extraordinary facts
in natural history, it was said to be the spot where
Hercules, fighting with the sons of Neptune, and
being in want of weapons, was supplied from hea-
ven by a shower of stones : hence it was called
Campus Herculeus.

This plain is on the east side of the Rhone,
between Salon and Arles: it is of a triangular
form, about twenty square leagues in extent, and
is covered almost entirely with quartzy gravel.
This immense collection of gravel has been sup-

posed by some to have been brought down by the Durance from the Alps of Dauphiné; by others it has been ascribed to the Rhone; and by many to the sea, as being a work too great for any river. The explanation mentioned above, § 105, namely, that the loose gravel on the plain arises from the decomposition of a great stratum of pudding-stone, which is the basis of the whole, is the opinion of Saussure, and is founded on his own observations. *

335. The theories that have been contrived for explaining the phenomena of the plain of Crau, afford an instance of the necessity of generalizing our observations before we can explain a particular appearance : in other words, they prove the truth of Lord Bacon's maxim, That the explanation of a phenomenon should not be sought for from the study of that phenomenon alone, but from the comparison of it with others. One of the theories of this plain is, that the breccia, which is the base of it, is formed from the consolidation of the loose gravel of the plain, by water percolating through it, and carrying some cementing substance along with it, or some *lapidific juice*, as it is called. And indeed, whether the gravel is formed from the brec-

---

* See Voyages aux Alpes, Tom. III. § 1592 et 1597. See also on this subject a Memoir by Lamanon, Journal de Physique, Tom. XXII. p. 477 ; and another by M. De Servieres, ibid. p. 270.

cia, or the breccia from the gravel, is a question
which probably could never be resolved by the mere
examination of the plain itself. But the question
is very soon decided, when we compare what is observed here with other appearances in the natural
history of the earth's surface, and consider how
much more frequent the decomposition of solids is,
than their reconsolidation, in any place above the
level of the sea.

336. The argument for the decomposition of
stony substances which is afforded by the state of
this singular plain, may be confirmed by the appearances observed in many extensive tracts of land
all over the world, and especially in some parts of
Great Britain. The road to Exeter from Taunton
Dean, between the latter and Honiton, passes over
a large heath or down, considerably elevated above
the plain of Taunton. The rock which is the base
of this heath, as far as can be discovered, is limestone, and over the surface of it large flints, in the
form of gravel, are very thickly spread. There is
no higher ground in the neighbourhood from which
this gravel can be supposed to have come, nor any
stream that can have carried it, so that no explanation of it remains, but that it is formed of the flints
contained in beds of limestone, which are now worn
away. The flints on the heath are precisely of the
kind found in limestone ; many of them are not
much worn, and cannot have travelled far from the

rock in which they were originally contained. It seems certain, therefore, that they are the *débris* of limestone strata, now entirely decomposed, that once lay above the strata which at present form the base of this elevated plain, and probably covered them to a considerable height. This explanation carries the greater probability with it, that any other way of accounting for the fact in question, as the travelling of the gravel from higher grounds, or the immersion of the surface under the sea, will imply changes in the face of the country, incomparably greater than are here supposed. Our hypothesis seems to give the *minimum* of all the kinds of change that can possibly account for the phenomenon.

337. The same remarks may be made on the high plain of Halldon, which the road passes over in going from Exeter to the westward. The flints there are disseminated over the surface as thickly as in the other instance, and can be explained only on the same supposition.

Again, in the interior of England, beginning from about Worcester and Birmingham, and proceeding north-east through Warwickshire, Leicestershire, Nottinghamshire, as far as the south of Yorkshire, a particular species of highly indurated gravel, formed of granulated quartz, is found every where in great abundance. This same gravel extends to the west and north-west, as far as Ashburn in Derbyshire, and perhaps still farther to the north. The quantity of it about Birmingham is very remarkable, as well as in many other places ; and the phenomenon is the more surprising, that no rock of the same sort is seen in its native place. It is such gravel as might be expected in a mountainous country, in Scotland, for instance, or in Switzerland, but not at all in the fertile and secondary plains of England.

This enigma is explained, however, when it is observed, that the basis of the whole tract just described is a red sandstone, often containing in it a hard quartzy gravel, perfectly similar to that which has just been mentioned. From the dissolution of beds of this sandstone, which formerly covered the present, there can be no doubt that this gravel is derived. But, as the gravel is in general thinly dispersed through the sandstone, and abounds only in some of its layers, it should therefore seem, that a vast body of strata must have been worn away and decomposed, before such quantities of gravel as now exist in the soil could have been let loose.

338. I have said, that a rock capable of affording such gravel as this, is not to be found in the tract of country just mentioned. This, however, is not strictly true ; for in Worcestershire, between Bromesgrove and Birmingham, about seven miles from the latter, a rock is found consisting of indu-

rated strata, greatly elevated, and without doubt
primitive, from the detritus of which such gravel as
we are now speaking of might be produced. These
strata seem to rise up from under the secondary,
where they are intersected by the road ; and, for as
much as appears, are not of great thickness, so that
they cannot have afforded the materials of this gravel directly, though they may have done so indirect-
ly, or through the medium of the red sandstone ;
that is to say, a primary rock of which they are the
remains, may have afforded materials for the gravel
in the sandstone ; and this sandstone may in its
turn have afforded the materials of the present soil,
and particularly the gravel contained in it.

339. Pudding-stones being very liable to decomposition, have probably, in most countries, afforded
a large proportion of the loose gravel now found in
the soil.   The mountains, or at least hills, of this
rock, which are found in many places, prove the
great extent of such decomposition.   Mount Rigi,
for instance, on the side of the Lake of Lucerne, is
entirely of pudding-stone, and is 742 toises in
height, measured from the level of the lake.   By
the descriptions given of it, as well as of other hills
of the same kind in Switzerland, we may, without
due attention, be led to suppose that they are entirely formed of loose gravel.   Even M. Saussure's
description is chargeable with this fault, though,
when attended to, it will be found to contain a suf-

ficient proof, that this hill is composed of real pudding-stone. *   The nature of the thing also, would
be sufficient to convince us, that a hill, more than
4000 feet in height, could not consist of loose and
unconsolidated materials.

If, then, we regard Mount Rigi as the remains
of a body of pudding-stone strata, we must conclude,
that these strata were originally more extensive,
and the adjacent valleys and plains will serve, in
some degree, to measure the quantity of them which
time has destroyed.

340. If the theory of unstratified mountains,
namely those of whinstone, porphyry, and granite,
be admitted as laid down above, it will furnish a
measure of the destruction which has taken place
in the stratified rocks, and of the vast depredations
which have been made upon them since they were
raised up from the bottom of the sea.   Like every
other measure, however, of wasting, by a thing that
is itself subject to waste, it can only give a *minimum*, or a limit which the quantity wasted must
necessarily exceed.

The abrupt face of a whinstone rock must be
understood as an evidence, that some body of strata
which supported it when fluid, remained in contact
with it, when it was become solid ; and if this part
of the mould in which the whinstone was cast, has

---

* Voyages aux Alpes, Tom. IV. § 1941.

disappeared, it must generally be ascribed to the operation of waste and decomposition. Such a face, for instance, as that which Salisbury Crag presents to the west, viz. a perpendicular wall of whinstone, about ninety feet high, raised on a body of sandstone strata of the height of about 300 feet, can have been produced only by having been abutted against some stratified rock, equally abrupt, and of the same elevation with itself. Of this rock no part remains.

The basaltic rock of Edinburgh Castle is nearly in the same state. Its perpendicular sides on the south, west, and north, are now disengaged from the strata by which they were once encompassed.

341. The granite mountains also, where they are quite unstratified, give rise to the same conclusion. Those central chains which we find in so many instances towering above the schistus which cover their sides, have probably been once completely enveloped by the latter ; and, on this supposition, an estimate may sometimes be formed of the original height of such mountains.

In these estimations, however, some uncertainty must arise, from our being unable to distinguish between the effects which are to be ascribed to the fracture and dislocation that took place when the compound body of stratified and unstratified rocks was raised up from the bottom of the sea, and the effects produced by the subsequent waste and decomposition at the surface. In this, as in many other instances, we are not always able to separate between the original inequalities of the surface, and those which wearing has produced.

342. It would be important to ascertain the rate at which the elevation of mountains decreases, and this is what we may perhaps expect to be accomplished, by the progress of geological science, and the multiplying of accurate observations. It has been supposed, that the Pyrenees diminish about ten inches in a century ; but what confidence is to be put in this estimate, I am unable to determine. *

A very unequivocal mark of the degradation of mountains is often to be met with in the heaps of loose stones found on their tops. These stones, it is obvious, cannot have come from any other place by natural means, and they are accordingly always sharp and angular, and have none of the characters of transported rocks. They are said sometimes to have been brought by men's hands ; but this is highly improbable, their quantity is often so considerable, and the difficulty of transportation so great. Where any purpose was to be served by heaping them together, men have availed themselves of the stones that they found ready prepared on the summit, and have constructed from them cairns, which

* Essai sur la Mineralogie des Pyrenées, p. 87.

have served as signals, useful in their pastoral, and sometimes in their military occupations.

NOTE XVIII. § 112.

Transportation of Stones, &c.

343. Nature supplies the means of tracing with considerable certainty the migration of fossil bodies on the surface of the earth, as only the more indurated stones, and those most strongly characterized, can endure the accidents that must befal them in travelling to a distance from their native place.

It is a fact very generally observed, that where the valleys among primitive mountains open into large plains, the gravel of those plains consists of stones, evidently derived from the mountains. The nearer that any spot is to the mountains, the larger are the gravel stones, and the less rounded is their figure ; and, as the distance increases, this gravel, which often forms a stratum nearly level, is covered with a thicker bed of earth or vegetable soil. This progression has particularly been observed in the valleys of Piémont and the plains of Lombardy, where a bed of gravel forms the basis of the soil, from the foot of the Alps to the shores of the Hadriatic. *

---

* Voyages aux Alpes, Tom. III. § 1315.

We may collect from Guettard, that a similar gradation is found in the gravel and earth which cover the great plain of Poland, from Mount Krapack to the Baltic. * The reason of this gradation is evident ; the farther the stones have travelled, and the more rubbing they have endured, the smaller they grow, the more regular is the figure they assume, and the greater the quantity of that finer detritus which constitutes the soil. The washing of the rains and rivers is here obvious ; and each of the three quantities just mentioned, if not directly proportional to the distance which the stones have migrated from their native place, may be said, in the language of geometry, to be at least proportional to a certain function of that distance.

344. The immense quantity of *cailloux roulés*, or rounded gravel, collected in the immediate vicinity of mountainous tracts, has led some geologists to suppose the existence of ancient currents, which descended from the mountains, in a quantity, and with a *momentum*, of which there is no example in the present state of the world. Thus Saussure imagines, that the hill of Superga, near Turin, which is formed of gravel, can only be explained by supposing such currents as are just mentioned, or what he terms a *débâcle*, to have taken place at

* Mém. Acad. des Sciences, 1762, p. 234, 293, &c.

some former period. * If, however, we ascribe to
the mountains a magnitude and elevation vastly
greater than that which they now possess ; if we
regard the valleys between them as cut out by the
rivers and torrents from an immense rampart of
solid rock, neither materials sufficiently great, nor
agents sufficiently powerful, will appear to be wanting, for collecting bodies of gravel and other loose
materials, equal to any that are found on the surface of the earth.   The necessity of introducing a
débâcle, or any other unknown agent, to account
for the transportation of fossils, seems to arise from
underrating the effects of action long continued,
and not limited by such short periods as circum-
scribe the works, and even the observations, of
men.

345. The supply of gravel and *cailloux roulés*,
for the plains extended at the feet of primitive
mountains, is doubtless in many cases much increased by the pudding-stone, interposed between
the secondary and the primary strata.   The beds of
pudding-stone contain gravel already formed on the
shores of continents, that ceased to exist before the
present were produced ; and the cement of this
gravel, yielding easily to the weather, allows the
stones included in it to be washed down by the
torrents, and scattered over the plains.   I know

---

* Voyages aux Alpes, Tom. III. § 1303.

not if the hill of Superga above mentioned, is not
in reality a mass of the pudding-stone which forms
the border of the Alps, and of which the materials
have suffered no transportation since the time of
their last consolidation.   This at least is certain,
that Saussure, notwithstanding his accuracy, has
sometimes confounded the loose gravel on the surface with that which is consolidated into rock ; an
inaccuracy which is to be charged, as I have elsewhere observed, rather against his system than himself.

346. The loose stones found on the sides of
hills, and the bottoms of valleys, when traced back
to their original place, point out with demonstrative evidence the great changes which have happened since the commencement of their journey ;
and in particular serve to show, that many valleys
which now deeply intersect the surface, had not
begun to be cut out when these stones were first
detached from their native rocks.   We know, for
instance, that stones under the influence of such
forces as we are now considering, cannot have first
descended from one ridge, and then ascended on
the side of an opposite ridge.   But the granite of
Mont Blanc has been found, as mentioned above,
on the sides of Jura, and even on the side of it
farthest from the Alps.   Now, in the present state
of the earth's surface, between the central chain of
the Alps, from which these pieces of granite must

have come, and the ridge of Mont Jura, besides many smaller valleys, there is the great valley of the Rhone, from the bottom of which, to the place where they now lie, is a height of not less than 3000 feet. Stones could not, by any force that we know of, be made to ascend over this height. We must therefore suppose, that when they travelled from Mont Blanc to Jura, this deep valley did not exist, but that such an uniform declivity, as water can run on with rapidity, extended from the one summit to the other. This supposition accords well with what has been already said concerning the recent formation of the Leman Lake, and of the present valley of the Rhone.

347. We can derive, in a matter of this sort, but little aid from calculation ; yet we may discover by it, whether our hypothesis transgresses materially against the laws of probability, and is inconsistent with physical principles already established. The horizontal distance from Mont Jura to the granite mountains, at the head of the Arve, may be accounted fifty geographic miles. Though we suppose Mont Blanc, and the rest of those mountains, to have been originally much higher than they are at present, the ridge of Jura must have been so likewise ; and though probably not by an equal quantity, yet it is the fairest way to suppose the difference of their height to have been nearly the same in former ages that it is at present,

and it may therefore be taken at 10,000 feet. The declivity of a plane from the top of Mont Jura to the top of Mont Blanc, would therefore be about one mile and three quarters in fifty, or one foot in thirty ; an inclination much greater than is necessary for water to run on, even with extreme rapidity, and more than sufficient to enable a river or a torrent to carry with it stones or fragments of rock, almost to any distance.

Saussure, in relating the fact that pieces of granite are found among the high passes near the summits of Mont Jura, alleges, that they are only found in spots from which the central chain of the Alps may be seen.   But it should seem that this coincidence is accidental, because, from whatever cause the transportation of these blocks has proceeded, the form of the mountains, especially of Mont Jura, must be too much changed to admit of the supposition, that the places of it from which Mont Blanc is now visible, are the same from which that mountain was visible when these stones were transported hither.   It may be, however, that the passes which now exist in Mont Jura are the remains of valleys or beds of torrents, which once flowed westward from the Alps ; and it is natural, that the fragments from the latter mountains should be found in the neighbourhood of those ancient water-tracks.

348. Saussure observed in another part of the

l

Alps, that where the Drance descends from the sides of Mont Velan and the Great St Bernard, to join the Rhone in the Valais, the valley it runs in lies between mountains of primary schistus, in which no granite appears, and yet that the bottom of this valley, toward its lower extremity, is for a considerable way covered with loose blocks of granite. * His familiar acquaintance with all the rocks of those mountains, led him immediately to suspect, that these stones came from the granite chain of Mont Blanc, which is westward of the Drance, and considerably higher than the intervening mountains. This conjecture was verified by the observations of one of his friends, who found the stones in question to agree exactly with a rock at the point of Ornex, the nearest part of the granite chain.

In the present state of the surface, however, the valley of Orsiere lies between the rocks of Ornex and the valley of the Drance, and would certainly have intercepted the granite blocks in their way from the one of these points to the other, if it had existed at the time when they were passing over that tract. The valley of Orsiere, therefore, was not formed, when the torrents, or the glaciers transported these fragments from their native place.

Mountainous countries, when carefully examined,

* Voyages aux Alpes, Tom. II. § 1022.

afford so many facts similar to the preceding, that
we should never have done were we to enumerate
all the instances in which they occur.   They lead
to conclusions of great use, if we would compare
the machinery which nature actually employs in
the transportation of rocks, with the largest fragments of rock which appear to have been removed,
at some former period, from their native place.

349. For the moving of large masses of rock, the
most powerful engines without doubt which nature
employs are the glaciers, those lakes or rivers of ice
which are formed in the highest valleys of the Alps,
and other mountains of the first order.   These
great masses are in perpetual motion, undermined
by the influx of heat from the earth, and impelled
down the declivities on which they rest by their
own enormous weight, together with that of the innumerable fragments of rock with which they are
loaded.   These fragments they gradually transport
to their utmost boundaries, where a formidable wall
ascertains the magnitude, and attests the force, of
the great engine by which it was erected.   The
immense quantity and size of the rocks thus transported, have been remarked with astonishment by
every observer, * and explain sufficiently how frag-

---

* The stones collected on the *Glacier de Miage,* when
Saussure visited it, were in such quantity as to conceal the
ice entirely.   Voyages aux Alpes, Tom. II. § 854.

ments of rock may be put in motion, even where there is but little declivity, and where the actual surface of the ground is considerably uneven. In this manner, before the valleys were cut out in the form they now are, and when the mountains were still more elevated, huge fragments of rock may have been carried to a great distance; and it is not wonderful, if these same masses, greatly diminished in size, and reduced to gravel or sand, have reached the shores, or even the bottom, of the ocean.

350. Next in force to the glaciers, the torrents are the most powerful instruments employed in the transportation of stones. These, when they descend from the sides of mountains, and even where the declivity of their course is not very great, produce effects which nothing but direct experience could render credible. The fragments of rock which oppose the torrent, are rendered specifically lighter by the fluid in which they are immersed, and lose by that means at least a third part of their weight: they are, at the same time, impelled by a force proportional to the square of the velocity with which the water rushes against them, and proportional also to the quantity of gravel and stones which it has already put in motion. Perhaps, after taking all these circumstances into computation, in the midst of a scene perfectly quiet and undisturbed, a philosopher might remain in doubt as to the power of torrents to move the enormous bodies of rock which

are seen in the bottom of the narrow valleys or deep glens of a mountainous country ; but his in- credulity, says an experienced traveller, will cease altogether, if he has been surprised by a storm in the midst of some Alpine region ; if he has seen the number and impetuosity of the cataracts which rush- ed down the sides of the mountains, and beheld the ruin which accompanied them ; and if, when the tempest was passed, he has viewed those meadows, which a few hours before were covered with ver- dure, now buried under heaps of stones, or over- whelmed by masses of liquid mud, and the sides of the mountains cut by deep ravines, where the track of the smallest rivulet was not before to be discovered. *

It is but rarely, however, even on occasions like these, that such vast masses of rock can be seen actually in motion, as are often found on the sur- face, apparently removed to a great distance from their native place. The magnitude of these is so great, in many instances, that their transportation cannot be explained without supposing, that the surface was very different when these transporta- tions took place from what it is at present ; that the elevation of the mountains was greater, and the ground smoother and more uniform, at least in some

---

* See an account of a thunder storm near Barrèges, in the Essai sur la Mineralogie des Pyrenées, p. 134.

directions. If these suppositions are admitted, and they are countenanced, as we have already seen, by almost every phenomenon in geology, the difficulties which present themselves here will not appear insurmountable.

351. One of the largest blocks of granite that we know of, is on the east side of the lake of Geneva, called Pierre de Gouté, about ten feet in height, with a horizontal section of fifteen by twenty. * Another block not far from it, and nearly of the same size, has some remains of schistus attached to it. These stones very much resemble those which have fallen from the Aiguilles, in the valley of Chamouni. The distance from their present situation to those Aiguilles is about thirty English miles, with many mountains and valleys at present interposed. By whatever means, therefore, these blocks were transported, their motion must have been over a surface of much more uniform declivity than the present. If the surface was without great inequalities, and its general declivity about one foot in thirty, as already computed, the glaciers, in the first place, and the torrents afterwards, may have served for the transportation even of these rocks.

352. Again, in the narrow vale or glen which separates the Great from the Little Saleve, the strata are all calcareous, but a great number of

* Voyages aux Alpes, Tom. I. § 308.

loose blocks of granite and primary schistus are scattered over the surface. A block of the former, near the lower end of the valley, is about the size of 1200 cubic feet. Two other large blocks of the same kind of stone rest on a base of horizontal limestone, elevated two or three feet above the rest of the surface. This elevation arises no doubt from the protection which the stones have afforded to the calcareous beds on which they lie, so that these beds do not wear away so fast as those which are fully exposed to the weather. But it is surely to take a very limited view of the operations on the surface, to suppose, with Saussure, that the parts of the calcareous rock under these stones has suffered no waste whatsoever, so that the stones remain now in the identical spot where they were placed by the great *débâcle* which brought them down from the high Alps. * For my part, I have no doubt that the Arve, which is still at no great distance, when it ran on a higher level, and in a line different from the present, aided by the glaciers and superior elevation of the mountains, was an engine sufficiently powerful for effecting the transportation of these stones.

353. These phenomena are not peculiar to the Alps, but prevail, in a greater or less degree, in the vicinity of all primary or granite mountains. In the island of Arran, a fragment of the same kind with

---

* Ibid. § 227.

that which constitutes the upper part of Goatfield, is found on the sea shore, at least three miles from the nearest granite rock, and with a bay of the sea intervening. Its dimensions are not far from those of the *pierre de Gouté*. In some former state of the granitic mountains in that island, the declivity from the top of Goatfield may have been very uniform, and more rapid than it is at present.

354. Besides glaciers and torrents, which have no doubt been the principal instruments in producing these changes, other causes may have occasionally operated. Large stones, when once detached, and resting on an inclined plane, from the effects of waste and decomposition, may advance horizontally, at the same time that they descend perpendicularly, and this will happen though they be not urged by any torrent, or any thing but their own weight; for the surface of the ground, as it wastes, remains higher under the stone, and for a little way round it, than at a greater distance, on account of the protection which it receives from the stone, as in the instances at Saleve, just mentioned. The stone itself also becomes rounded at the bottom; and thus the surface in contact with the ground is diminished in extent, and the two surfaces rendered convex towards one another. It must therefore happen, that the support, continually weakening, will at length give way, and the stone incline or roll toward the lower side, and may even roll consider-

ably, if its centre of gravity has been high above its point of support, and if its surface has had much convexity : Thus the horizontal may very far exceed the perpendicular motion ; and, in the course of ages, the stone may travel to a great distance. A stone, however, which travels in this manner, must diminish as it proceeds, and must have been much greater in the beginning than it is at present.

355. This kind of motion may be aided by particular circumstances. When a stone rests on an inclined plane, so as to be in a state not very remote from equilibrium, if a part be taken away from the upper side, the equilibrium will be lost, and the stone will thereby be put in motion. That stones which lie on other stones, may, by wearing, be brought very near an equilibrium, is proved by what are called rocking-stones, or in Cornwall Logan stones, which have sometimes been mistaken for works of art ; but are certainly nothing else than stones, which have been subjected to the universal law of wasting and decay, in such peculiar circumstances, as nearly to bring about an equilibrium of that stable kind, which, when slightly disturbed, re-establishes itself. * The Logan stone at the

* I do not presume so far as to say, that all rocking-stones are produced by natural means : I have not sufficient information to justify that assertion ; but the great size of that at the Land's End, its elevated position, and the ap-

Land's End, is a mass of granite, weighing more than sixty tons, resting on a rock of granite, of considerable height, and close on the sea shore. The two stones touch but in a small spot, their surfaces being considerably convex towards one another. The uppermost is so nearly in an equilibrium, that it can be made to vibrate by the strength of a man, though to overset it entirely would require a vast force. This arises from the centre of gravity of the stone being somewhat lower than the centre of curvature of that part of it on which it has a tendency to roll ; the consequence of which is, that any motion impressed on the stone, forces its centre of gravity to rise, (though not very considerably,) by which means it returns whenever the force is removed, and vibrates backward and forward, till it is reduced to rest. Were it required to remove the stone from its place, it might be most easily done, by cutting off a part from one side, or blowing it away by gunpowder ; the stone would then lose its balance, would tum-

---

proaches toward something of the same kind which are to be seen in other parts of that shore, prove that it is no work of art. They who ascribe it to the Druids, do not consider the rapidity with which the Cornish granite wastes, nor think how improbable it is, that the conditions necessary to a rocking-stone, whether produced by nature or art, should have remained the same for sixteen or seventeen hundred years.

ble from its pedestal, and might roll to a considerable distance. Now, what art is here supposed to
perform, nature herself in time will probably effect.
If the waste on one side of this great mass shall exceed that on the opposite in more than a certain
proportion, and it is not likely that that proportion
will be always maintained, the equilibrium of the
Logan stone will be subverted, never to return.
Thus we perceive how motion may be produced by
the combined action of the decomposition and gravitation of large masses of rock.

356. Besides the gradual waste to which stones
exposed to the atmosphere are necessarily subject,
those of a great size appear to be liable to splitting,
and dividing into large portions, no doubt from
their weight. This may be observed in almost all
stones that happen to be in such circumstances as
we are now considering ; and from this cause the
subversion of their balance may be more sudden,
and of greater amount, than could be expected from
their gradual decay.

Thus, if to the wasting of a stone at the bottom,
we add the accidents that may befal it in the wasting of its sides, we see at least the physical possibility of detached stones being put in motion, merely by their own weight. It is indeed remarkable,
that some of the largest of these stones rest on
very narrow bases. Those at the foot of Saleve
touch the ground only in a few points : The

Boulder stone of Borrowdale is supported on a narrow ridge like the keel of a ship, and is prevented from tumbling by a stone or two, that serve as a kind of shores to prop it up. Very unexpected accidents sometimes happen to disturb the rest of such fragments of rock as have once migrated from their own place. Saussure mentions a great mass of *lapis ollaris*, * that lies detached on the side of a declivity in the valley of Urseren, in the canton of Uri. The people use this stone as a quarry, and are working it away on the upper side, in consequence of which it will probably be soon overset, and will roll to the bottom of the valley.

357. In many instances it cannot be doubted, that stones of the kind here referred to are the remains of masses or veins of whinstone or granite, now worn away, and that they have travelled but a very short way, or perhaps not at all, from their original place. Many of the large blocks of whinstone which we find in this country, sometimes single, and sometimes scattered in considerable abundance over a particular spot, are certainly to be referred to this cause. But the most remarkable examples of this sort are the stones found at the Cape of Good Hope, on the hill called Paarlberg, which takes its name from a chain of large

* Voyages aux Alpes, Tom. IV. § 1851.

round stones, like the pearls of a necklace, that
passes over the summit. Two of these, placed
near the highest point, are called the Pearl and
the Diamond, and were mentioned several years
ago in the Philosophical Transactions. *   From a
more recent account, these stones appear to be a
species of granite, though the hill on which they
lie is composed of sandstone strata. †   The Pearl
is a naked rock, that rises to the height of 400
feet above the summit of the hill ; the Diamond
is higher, but its base is less, and it is more inaccessible.

From the above stones forming a regular chain,
as well as from the immense size of the two largest, it is impossible to suppose that they have been
moved ; and it is infinitely more probable, that they
are parts of a granite vein, which runs across the
sandstone strata, and of which some parts have resisted the action of the weather, while the rest
have yielded to it.   The whole geological history
of this part of Africa seems highly interesting,
since, as far as can be collected from the accounts
of the ingenious traveller just mentioned, it consists of horizontal beds of sandstone or limestone,
resting immediately on granite, or on primary

---

* Vol. LXVIII. p. 102.

† Barrow's Travels into Southern Africa, p. 60.

schistus.   Loose blocks of granite are seen in great abundance at the foot of the Table Mountain, and along the sea shore.

———

358. The system which accounts for such phenomena as have been considered in this and some of the preceding notes, by the operation of a great deluge, or *débâcle*, as it is called, has been already mentioned.   In Dr Hutton's theory, nothing whatever is ascribed to such accidental and unknown causes ; and, though their existence is not absolutely denied, their effects, whatever they may have been, are alleged to be entirely obliterated, so that they can be referred to no other class but that of mere possibilities.   A minute discussion, however, of the question, Whether there are, on the surface of the earth, any effects that require the interposition of an extraordinary cause, would lead into a longer digression than is suited to this place. I shall briefly state what appear to be the principal objections to all such explanations of the phenomena of geology.

359. The general structure of valleys among mountains, is highly unfavourable to the notion that they were produced by any single great torrent, which swept over the surface of the earth. In some instances, valleys diverge, as it were from

a centre, in all directions. In others, they originate from a ridge, and proceed with equal depth
and extent on both sides of it, plainly indicating,
that the force which produced them was *nothing*,
or evanescent at the summit of that ridge, and increased on both sides, as the distance from the
ridge increased. The working of water collected
from the rains and the snows, and seeking its way
from a higher to a lower level, is the only cause we
know of, which is subject to this law.

360. Again, if we consider a valley as a space,
which perhaps with many windings and irregularities, has been hollowed out of the solid rock, it is
plain, that no force of water, suddenly applied,
could loosen and remove the great mass of stone
which has actually disappeared. The greatest column of water that could be brought to act against
such a mass, whatever be the velocity we ascribe to
it, could not break asunder and displace beds of
rock many leagues in length, and in continuity with
the rock on either side of them. The slow working of water, on the other hand, or the powers that
we see every day in action, are quite sufficient for
this effect, if time only is allowed them.

361. Some valleys are so particularly constructed, as to carry with them a still stronger refutation
of the existence of a *débâcle*. These are the longitudinal valleys, which have the openings by which
the water is discharged, not at one extremity, but

at the broadside.   Such is that on the east side of
Mont Blanc, deeply excavated on the confines of
the granite and schistus rock, and extending parallel to the beds of the latter, from the Col de la
Seigne to the Col de Ferret ; its opening is nearly
in the middle, from which the Dora issues, and
takes its course through a great valley, nearly at
right angles to the chain of the Alps, and to the
valley just mentioned. From the structure of these
valleys, Saussure has argued very justly against
Buffon's hypothesis, concerning the formation of
valleys by currents at the bottom of the sea. *   It
affords indeed a complete refutation of that hypothesis : and it affords one no less complete of the
system which Saussure himself seems on some occasions so much inclined to support.   For if it be
said, that this valley was cut out by the current of
a *débâcle*, that current must either have run in the
direction of the valley of Ferret, or in that of the
Dora, which issues from it.   If it had the direction of the first, it could not cut out the second ;
and if it had the direction of the second, it could
not cut out the first.   Besides, the force which excavated this valley must have been *nothing* at the
two extreme points, viz. at the Col de la Seigne
and the Col de Ferret, and must have increased
with the distance from each.   It can have been

---

* Voyages aux Alpes, Tom. II. § 920.

produced, therefore, only by the running of two streams in opposite directions, on a surface that was but slightly uneven, these streams at meeting taking a new direction, nearly at right angles to the former. A clearer proof could hardly be required than is afforded in this case, that what is now a deep valley was formerly solid rock, which the running of the waters has gradually worn away ; and that the waters, when they began to run, were on a level as high, at least, as the tops of those mountains by which the valley is bounded toward the lower side.

362. Longitudinal valleys, with the water bursting out transversely from their sides, like the preceding, are by no means confined to mountains of the first order. We have a very good example, though on a small scale, of a valley of this sort, within a few miles of Edinburgh. The Pentland Hills form a double ridge, separated by a small longitudinal valley, that runs from N. E. to S. W., the water of which issues from an opening almost in the middle, and directed towards the south. This, therefore, is not the work of any great torrent, which overwhelmed the country ; for no one direction, which it is possible to assign to such a torrent, will afford an explanation, both of the valley and its outlet. *

---

* In Scotland there is one valley, of a kind that I believe

363. They who maintain the existence of the
débâcle, will no doubt allege, that though these

---

is extremely rare in any part of the world, in accounting for
which, the hypothesis of a torrent or *débâcle* might, if any
where, be employed to advantage. This is the valley which
extends across the island, from Inverness to Fort William,
or from sea to sea, being open at both ends, and very little
elevated in the middle. It is nearly straight, and of a very
uniform breadth, except that towards each end it widens
considerably. The bottom, reckoning transversely, is flat,
without any gradual slope from the sides towards the middle. From the sides the mountains rise immediately, and
form two continued ridges of great height, like ramparts or
embankments on each side of a large fossé. A great part of
the bottom of this singular valley is occupied by lakes,
namely, Loch Ness, Loch Oich, and Loch Lochy. Its
length is about sixty-two miles, and the point of partition
from which the waters run different ways, viz. north-east to
the German Ocean, and south-west to the Atlantic, is between Loch Oich and Loch Lochy ; and, by the estimation
of the eye, I should hardly think that it is elevated more
than ten or fifteen feet above the surface of either lake. The
country on both sides is rugged and mountainous, and the
streams which descend from thence into the valley, either
fall directly into the lakes, or turn off almost at right angles
when they enter the valley. Though the bottom of this valley, therefore, is every where alluvial, with the exception,
perhaps, of a few rocks which appear at the surface, it is certainly not excavated by the rivers which now flow in it.
The direction of the valley, it is to be observed, is the same
with that of the vertical strata which compose the mountains
on either side.

valleys were not cut out by means of it, yet others may. But it must be recollected, that if some of

---

Here, then, we have a valley, not cut out by the working of any streams which now appear; and we may therefore make trial of the hypothesis of a *débâcle*. This, however, will afford us no assistance; because, if we suppose what is now hollow to have been once occupied by the same kind of rock which is on either side, no force of torrents can have suddenly loosened and removed from its place a body of such vast magnitude. A greater column of water, than one having for its base a transverse section of the valley, could not act against it, and this would have to overcome the cohesion and inertia of a column of rock of the same section, and of the length of sixty-two miles. It is not hazarding much to affirm, that no velocity which could be communicated to water, not even that which it could acquire by falling from an infinite height, could give to it a force in any degree adequate to this great effect.

The explanation of this valley, which appears to me the most probable, is the following. It will be shown hereafter, that there is good reason to suppose, that, in most parts of our island, the relative level of the sea and land has been in past ages considerably higher than it is at present. In such circumstances, this valley may have been under the surface of the sea, the highest part of it being scarcely 100 feet above that level at present. It may have been a kind of sound, therefore, or strait, which connected the German Sea with the Atlantic; and the strong currents, which, on account of the different times of high water in these two seas, must have run alternately up and down this strait, may have produced that flatness of the bottom, and straightness of the sides, and that widening at the extremities, which are men-

4

the greatest and deepest valleys on the face of the
earth, such as that just mentioned, on the east side
of Mont Blanc, are thus shown to be the work of
the daily wasting of the surface, what other inequalities can be great enough to require the interposition of a more powerful cause ? If a *dignus
vindice nodus* does not exist here, in what part of
the natural history of the earth is it likely to be
found ?

364. The large masses of rock so often met with
at a distance from their original place, are one of
the arguments used for the *débâcle*. It has, however, been shown, that, supposing a form of the
earth's surface considerably different from the present, especially, supposing the absence of the valleys which the rivers have gradually cut out, the
transportation of such stones is not impossible, even
by such powers as nature employs at present. Now,
without the supposition that the surface was more
continuous, and that its present inequalities did not
exist, no force of torrents, whatever their velocity
and magnitude may have been, could have produc-

---

tioned above. In this way, too, some difficulties are removed relative to Loch Ness, which is so deep as hardly to be
consistent with the indefinite length of the period of waste
that must be ascribed to the mountains on each side of it.
Its depth is said, where greatest, not to be less than 180
fathoms. According to this hypothesis, it may, at no very
distant period, have been a part of the bottom of the sea.

ed this transportation.   No force of water could
raise a stone like the *pierre de Gouté* from the bottom of a valley, to the top of a steep hill.   Indeed,
if we suppose a great fragment of rock to be hurried along on a horizontal or an inclined plane, by
the force of water, the moment it comes to a deep
valley, and has to rise up over an ascent of a certain steepness, it will remain at rest ; the water itself will lose its velocity, and the heavy bodies
which it carried with it will proceed no farther.
Thus, therefore, we have the following dilemma.
If the surface is not supposed to have had a certain
degree of uniformity in past times, a *débâcle* is insufficient for the transportation of stones : If it is
supposed to have had that uniformity, a *débâcle* is
unnecessary.

365. Another fact, which has been supposed favourable to the opinion of the action of great torrents at some former period, is, that in countries
like that round Edinburgh, where whinstone hills
rise up from among secondary strata, a remarkable
uniformity is observed in the direction of their
abrupt faces.   Thus, in the country just mentioned, the steep faces generally front the west, while,
in the opposite direction, the slope is gentle, and
the hills decline gradually into the plain.   Hence
it is supposed, that a torrent, sweeping from west
to east, has carried off the strata from the west side
of these hills, but, being obstructed by the whinstone

rock, has left the strata on the east side in their natural place.

But, besides that no force which can ever be ascribed to a torrent could have removed at once bodies of strata 300 or 400 feet, nay even 800 or
1000 in thickness, which must have been the case
if this were the true explanation of the fact, there is
a circumstance which may perhaps enable us to explain these phenomena without the assistance of any
extraordinary cause. The secondary strata in which
the whinstone hills are found in this part of Scotland, are not horizontal, but rise or *head* towards
the west, dipping towards the east. The side,
therefore, of the whinstone hills which is precipitous, is the same with that towards which the strata
rise. Now, from the manner in which these hills
are supposed to have been elevated, the strata are
likely to have been most broken and shattered towards that side, while, on the opposite, they had
the support of the whinstone rock. They would become a prey, therefore, more easily to the common
causes of erosion and waste on the upper side than
on the lower. The streams that flowed from the
higher grounds would wear them on the former
most readily ; and the action of these streams
would be resisted by the superior hardness of the
whinstone, just as the great torrent of the *débâcle*
is supposed to have been.

It should also be observed, that this fact of the

uniform direction of the abrupt faces of mountains, is often too hastily generalized. In primitive countries, it is no farther observed than by the steep faces of the mountains being most frequently turned toward the central chain. In Scotland, as soon as you leave the flat country, and enter the Highlands, the scarps of the hills face indiscriminately all the points of the compass, and are directed as often to the east as to the west.

366. Where the strata are nearly horizontal, they afford the most distinct information concerning the direction and progress of the wasting of the land. The inclined position of the strata, which in all other cases must enter for so much into our estimate of the causes which have produced the present inequality of the earth's surface, disappears there entirely ; and the whole of that inequality is to be ascribed to the operations at the surface, whether they have been sudden or gradual. A very important fact from a country of this sort, is related by Barrow, in his Travels into Southern Africa. The mountains about the Cape of Good Hope, and as far to the north as that ingenious traveller prosecuted his journey, are chiefly of horizontal strata of sandstone and limestone, exhibiting the appearance, on their abrupt sides, of regular layers of masonry, of towers, fortifications, &c. Now, among all these mountains, he observed, that the high or steep sides look constantly down the rivers,

while the sloping or inclined sides have just the opposite direction.   When, in travelling northward, he passed the line of partition, where the waters from running south take their direction to the north, he found, that the gradual slope, which had hitherto been turned to the north, was now turned to the south : The abrupt aspect of the mountains, in like manner, from facing the south, was directed to the north ; so that, in both cases, the hills turned their backs on the line of greatest elevation. *

It is evident, therefore, that the form of this land has been determined by the slow working of the streams.   The causes which produced the effects here described, began their action from the line of greatest elevation, and extended it from thence on both sides, in opposite directions.   This is the most precise character that can mark the alluvial operations, and distinguish them from the overwhelming power of a great *débâcle*.

367. Lastly, If there were any where a hill, or any large mass composed of broken and shapeless stones, thrown together like rubbish, and neither worked into gravel nor disposed with any regularity, we must ascribe it to some other cause than the ordinary *detritus* and wasting of the land.   This, however, has never yet occurred ; and it seems

---

* Barrow's Travels into Southern Africa, p. 245.

best to wait till the phenomenon is observed, before we seek for the explanation of it.

368. These arguments appear to me conclusive against the necessity of supposing the action of sudden and irregular causes on the surface of the earth. In this, however, I am perhaps deceived : neither Pallas, nor Saussure, nor Dolomieu, nor any other author who has espoused the hypothesis of such causes, has explained his notions with any precision ; on the contrary, they have all spoken with such reserve and mystery, as seemed to betray the weakness, but may have concealed the strength of their cause. I have therefore been combating an enemy, that was in some respects unknown ; and I may have supposed him dislodged, only because I could not penetrate to his strongholds. The question, however, is likely soon to assume a more determinate form. A zealous friend of Dr Hutton's theory, has lately * declared his approbation of the hypothesis which has here been represented as so adverse to that theory ; and, from his ability and vigour of research, it is likely to receive every improvement of which it is susceptible.

---

* Trans. Royal Society Edin. Vol. V. p. 68.

Note xix. § 117.

Transportation of Materials by the Sea.

369. The existence of the great and extensive operations, by which the spoils of the land are carried all over the ocean, and spread out on the bottom of it, may be supposed to require some further elucidation. We must attend, therefore, to the following circumstances.

When the detritus of the land is delivered by the rivers into the sea, the heaviest parts are deposited first, and the lighter are carried to a greater distance from the shore. The accumulation of matter which would be made in this manner on the coast, is prevented by the farther operation of the tides and currents, in consequence of which the substances deposited continue to be worn away, and are gradually removed farther from the land. The reality of this operation is certain ; for otherwise we should have on the sea shore a constant and unlimited accumulation of sand and gravel, which, being perpetually brought down from the land, would continually increase on the shore, if nature did not employ some machinery for removing the advanced part into the sea, in proportion to the supply from behind.

The constant agitation of the waters, and the declivity of the bottom, are no doubt the causes of this gradual and widely extended deposition. A soft mass of alluvial deposit, having its pores filled with water, and being subject to the vibrations of a superincumbent fluid, will yield to the pressure of that fluid on the side of the least resistance, that is, on the side toward the sea, and thus will be gradually extended more and more over the bottom.   This will happen not only to the finer parts of the detritus, but even to the grosser, such as sand and gravel.   For suppose that a body of gravel rests on a plane somewhat inclined, at the same time that it is covered with water to a considerable depth, that water being subject not only to moderate reciprocations, but also to such violent agitation as we see occasionally communicated to the waters of the ocean ; the gravel, being rendered lighter by its immersion in the water, and on that account more moveable, will, when the undulations are considerable, be alternately heaved up and let down again.   Now, at each time that it is heaved up, however small the space may be, it must be somewhat accelerated in its descent, and will hardly settle on the same point where it rested before.   Thus it will gain a little ground at each undulation, and will slowly make its way towards the depths of the ocean, or to the lowest situation it can reach.   This, as far as we may presume to

follow a progress which is not the subject of immediate observation, is one of the great means by
which loose materials of every kind are transported
to a great distance, and spread out in beds at the
bottom of the ocean.

370. The lighter parts are more easily carried
to great distances, being actually suspended in the
water, by which they are very gradually and slowly deposited. A remarkable proof of this is furnished from an observation made by Lord Mul-
grave, in his voyage to the North Pole. In the
latitude of 65° nearly, and about 250 miles distant
from the nearest land, which was the coast of Norway, he sounded with a line of 683 fathoms, or
4098 feet; and the lead, when it struck the
ground, sunk in a soft blue clay to the depth of ten
feet. * The tenuity and fineness of the mud,
which allowed the lead to sink so deep into it,
must have resulted from a deposition of the lighter
kinds of earth, which being suspended in the water, had been carried to a great distance, and were
now without doubt forming a regular stratum at
the bottom of the sea.

371. The quantity of detritus brought down by
the rivers, and distributed in this manner over the
bottom of the sea, is so great, that several narrow
seas have been thereby rendered sensibly shallower.

---

* Phipps's Voyage, p. 74, 141.

The Baltic has been computed to decrease in depth
at the rate of forty inches in a hundred years. The
Yellow Sea, which is a large gulf contained between
the coast of China and the peninsula of Corea, receives so much mud from the great rivers that run
into it, that it takes its colour, as well as its name,
from that circumstance; and the European mariners,
who have lately navigated it, observed, that the
mud was drawn up by the ships, so as to be visible
in their wake to a considerable distance. *   Com-
putations have been made of the time that it will
require to fill up this gulf, and to withdraw it entirely from the dominion of the ocean : but the
data are not sufficiently exact to afford any precise
result, and are no doubt particularly defective from
this cause, that much of the earth carried into the
gulf by the rivers, must be carried out of it by
the currents and tides, and the finer parts wafted
probably to great distances in the Pacific Ocean. †
The mere attempt, however, towards such a computation, shows how evident the progress of filling

* Staunton's Account of the Embassy to China, Vol. I. p.
448.

† Pérouse, in sailing along the coast of China, from For-
mosa to the strait between Corea and Japan, though generally fifty or sixty leagues from the land, had soundings at
the depth of forty-five fathoms, and sometimes at that of
twenty-two.   Atlas du Voyage de la Pérouse, No. 43.

up is to every attentive observer; and, though it may not ascertain the measure, it sufficiently declares the reality of the operations, by which the waste of the present continents is made subservient to the formation of new land.

372. Sandbanks, such as abound in the German Ocean, to whatever they owe their origin, are certainly modified, and their form determined, by the tides and currents. Without the operation of these last, banks of loose sand and mud could hardly preserve their form, and remain intersected by many narrow channels. The formation of the banks on the coast of Holland, and even of the Dogger Bank itself, has been ascribed to the meeting of tides, by which a state of tranquillity is produced in the waters, and of consequence a more copious deposition of their mud. Even the great bank of Newfoundland seems to be determined in its extent by the action of the Gulf stream. In the North Sea, the current which sets out of the Baltic, has evidently determined the shape of the sandbanks opposite to the coast of Norway, and produced a circular sweep in them, of which it is impossible to mistake the cause.

In proof of the action here ascribed to the waters of the sea, in transporting materials to an unlimited extent, we may add the well known observation, that the stones brought up by the lead from the bottom of the sea, are generally round and polished, hardly ever sharp and angular. This could never happen to stones that were not subject to perpetual attrition.

373. Currents are no doubt the great agents in diffusing the detritus of the land over the bottom of the sea. These have been long known to exist ; but it is only since the later improvements in navigation, that they have been understood to constitute a system of great permanence, regularity, and extent, connected with the trade winds, and other circumstances in the natural history of the globe. The Gulf stream was many years since observed to transport the water, and the temperature of the tropical regions into the climates of the north ; and we are indebted to the researches of Major Rennell, for the knowledge of a great system of currents, of which it is only a part. That geographer, who is so eminent for enriching the details of his science with the most interesting facts in history or in physics, has shown, that along the eastern coast of Africa, from about the mouth of the Red Sea, a current fifty leagues in breadth sets continually towards the southwest. * It doubles the Cape of Good Hope, runs from thence northwest, preserving on the whole the direction of the coast, but reaching so far into the ocean, that, about the parallel of St Helena, its breadth exceeds 1000 miles.

---

* Geography of Herodotus, p. 672.

From thence, as it approaches the line, its direction is more nearly east; and meeting in the parallel of 3° north, with a current which has come along the western coast of Africa from the north, the two united stretch across the Atlantic, in a line somewhat south of west, and in a very wide and rapid stream. This stream meets the American land at Cape St Roque, where it is joined by another coming up along the eastern shore of that continent, and directed towards the north. They proceed northward together till they enter the Gulf of Florida, from which being as it were reflected, they form the Gulf stream, passing along the coast of North America, and stretching across the Atlantic to the British Isles. From thence the current turns to the south, and, proceeding down the coast of Spain and Africa, meets the stream ascending from the south, as already described, and thus continues in perpetual circulation. The velocity of these currents is not less remarkable than their extent. At the Cape of Good Hope, the rate is thirty nautical miles in twenty four hours; in some places forty five; and under the line seventy seven. When the Gulf stream issues from the Straits of Bahama, it runs at the rate of four miles an hour, and proceeds to the distance of 1800 miles, before its velocity is reduced to half that quantity. In the parallel of 38°, near 1000 miles from the above strait, the water of the stream has been found ten degrees warmer than the air.

374. The course of the Gulf stream is so fixed and regular, that nuts and plants from the West Indies are annually thrown ashore on the Western Islands of Scotland. The mast of a man of war, burnt at Jamaica, was driven several months afterwards on the Hebrides, * after performing a voyage of more than 4000 miles, under the direction of a current, which, in the midst of the ocean, maintains its course as steadily as a river does upon the land.

The great system of currents thus traced through the Atlantic, has no doubt phenomena corresponding to it in the Indian and Pacific Oceans, which the industry of future navigators may discover. The whole appears to be connected with the trade winds, the figure of our continents, the temperature of the seas themselves, and perhaps with some inequalities in the structure of the globe. The disturbance produced by these causes in the equilibrium of the sea, probably reaches to the very bottom of it, and gives rise to those counter currents, which have sometimes been discovered at great depths under the surface. †

The great transportation of materials that must

---

* Pennant's Arctic Zoology, Introd. p. 70.

† Histoire Naturelle de Buffon, Supplément, Tom. IX. p. 479. 8vo.

result from the action of these combined currents is obvious, and serves not a little to diminish our wonder, at finding the productions of one climate so frequently included among the fossils of another. Amid all the revolutions of the globe, the economy of nature has been uniform, in this respect, as well as in so many others, and her laws are the only thing that have resisted the general movement. The rivers and the rocks, the seas and the continents, have been changed in all their parts ; but the laws which direct those changes, and the rules to which they are subject, have remained invariably the same.

375. Objections have been made to that translation of materials by the waters of the ocean which is supposed in this theory, particularly by Mr Kirwan, in his Geological Essays ; and, though I might perhaps content myself with the remark already made, that the Neptunian system involves suppositions concerning the transportation of solid bodies by the sea, in the early ages of the world, as wonderful as those which, according to our theory, are common to all ages, I am unwilling to remain satisfied with a mere *argumentum ad hominem*, where the fallacy of the reasoning is so easily detected.

376. One of Mr Kirwan's objections to the deposition of materials at the bottom of the sea, is thus stated : " Frisi has remarked, in his mathematical discourses, that if any considerable mass of

matter were accumulated in the interior of the
ocean, the diurnal motion of the globe would be
disturbed, and consequently it would be perceptible ; a phenomenon, however, of which no history
or tradition gives any account." *

The appeal made here to Frisi is singularly un-
fortunate, as that philosopher has demonstrated the
very contrary of Mr Kirwan's position, and has
proved, that the disturbance given to the diurnal
motion by the causes here referred to may be real,
but cannot be perceptible. Having investigated a
formula expressing the law which all such disturbances must necessarily observe, he concludes, " Hâc
autem formulâ manifestum fiet, ex iis omnibus va-
riationibus quæ in terrestri superficie observari so-
lent, montium et collium abrasione, dilapsu corpo-
rum ponderosiorum in inferiores telluris sinus, nul-
lam oriri posse variationem *sensibilem* diurni motûs. Nam si statuamus data aliqua annorum pe-
riodo terrestrem superficiem ad duos usque pedes
abradi undique, eam vero materiæ quantitatem ad
profunditatem pedum 1000 dilabi ; erit omne quod
inde orietur incrementum velocitatis diurni motûs

$$\frac{30000}{(19638051)^2}=\frac{1}{12855068184}.\text{" }\dagger$$

Here, it is evident, that Frisi admits those very

---

* Geol. Essays, p. 441.
† Frisii Opera, Tom. III. p. 269.

changes on the surface which we are contending for, and shows, that their tendency is to accelerate the earth's diurnal motion, but, by a quantity so small, that, in a space of time amounting at least to 200 years, the increase of the diurnal motion would only be such a part of the whole as the preceding fraction is of unity. *

---

* The time requisite for taking away by waste and erosion two feet from the surface of all our continents, and depositing it at the bottom of the sea, cannot be reckoned less than 200 years. The fraction $\frac{1}{1283500081084}$, reduced to parts of a day, is $\frac{1}{148554}$ of a second; so that it would require 200 years to shorten the length of the day, by the above fraction of a second; and therefore it would require 148554 times 200 years, or 29710800 years, to diminish it an entire second. The accumulated effect, however, of all the diminutions during that period, would amount to much more: and if we had any perfectly uniform standard to compare the motion of the earth with, its difference from that standard would increase as the squares of the time, and the total acceleration would amount to one second in 77080 years. Whatever relation this bears to the age of the globe itself, it exceeds more than ten times the age of any historical record.

Though Frisi concludes, as is stated here, that the acceleration produced in the diurnal motion of the earth, is far too inconsiderable to become the object of astronomical observation, he makes a supposition difficult to be reconciled with this conclusion, namely, that the acceleration has had a sensible effect on the figure of the earth, or rather of the sea, having increased the centrifugal force, and thereby accumulated the waters under the equator, in the present,

377. The instance just given may serve as one of many, to show what confidence is to be placed in that indigested mass of facts and quotations which Mr Kirwan, without discrimination, and without discussion, has brought together from all quarters. He has no intention, I believe, to deceive his readers; but we may judge, from this specimen, of the precautions he has taken against being deceived himself.

In some respects, the result of Frisi's investigation must be considered as imperfect. If there were no relative motion in the parts of our globe, but that by which things descend from a higher to a lower level, a continual acceleration of its rotation, though extremely slow, would take place, as above computed. But as, in the interior of the earth, there are undoubtedly motions of a tendency opposite to those on the surface, and directed from the centre towards the circumference, they must

---

more than in former ages. Such an accumulation, he thinks agreeable to certain appearances that have been observed respecting the ancient level of the sea. These appearances will be afterwards considered: it is sufficient to remark here, that though the fraction, expressing the increment of the centrifugal force, must be double that which expresses the acceleration, it must be too small to have any perceptible effect in elevating the sea, except after an immense interval of time; and the compensations which arise from other causes, probably must prevent it from becoming sensible in any length of time whatsoever.

produce a retardation in the diurnal revolution ; and from this must arise an inequality, not uniformly progressive in the same direction, but periodical, and confined within certain limits, as the causes are by which it is produced. *

---

* Even in the descent of bodies from a higher to a lower level at the surface of the earth, the whole tendency is not to increase the velocity of the earth's rotation, and many compensations take place, which, when the matter is considered only in general, are necessarily overlooked. This will appear evident, if we reflect, that it is not simply the approach of a body towards the centre of the earth, or its removal from that centre, which tends to disturb the rotation of the earth ; but its approach to the axis of the earth, or its removal from that axis. The velocity with which a particle of matter revolves, whether on the surface, or in the interior of the globe, is proportional to its distance from the axis of rotation ; and therefore, when a body comes nearer to the axis, it loses a part of the motion which it had before ; which part, of consequence, is communicated to the whole mass of the earth, and therefore tends to increase the velocity with which it revolves. The contrary happens when a body recedes from the axis ; for it then receives an addition to its velocity, which, of course, is taken away from the rotatory motion of the earth.

Hence, bodies moving in a horizontal plane, may increase or diminish the swiftness of the diurnal motion, according as they move towards the poles or towards the equator ; and those which descend from a higher to a lower level, disturb the earth's rotation, much more in consequence of their horizontal, than of their perpendicular motion. The Ganges, for instance, though its source is probably elevated no less than

378. Mr Kirwan's second objection is founded
on the misapprehension of a well known fact in the

7000 feet above the level of the sea, tends to retard the
earth's rotation, by bringing its waters, and the mud contained in them, from the parallel of 31⁰ to that of 22⁰, and
so increasing their distance from the earth's axis by more
than $\frac{1}{13}$th part. Had the Ganges flowed towards the north,
as the Nile does, its effect would have been just the contrary.

In the same manner, a stone descending from the top of a
mountain, may accelerate or retard the earth's rotation, according to the direction in which it descends. If it descend
on the side of the elevated pole, it will then produce acceleration, because its distance from the axis will be diminished:
but if it descend on the side of the depressed pole, and if the
direction in which it is moved, be over a line less inclined,
than a line drawn from the same point to the depressed pole,
it will then produce a retardation, because its distance from
the axis will be increased.

Let us suppose, for example, that the top of Mont Blanc
is in latitude 45⁰ 49′, and that its height is 2450 toises above
the level of the sea. The point at which a line drawn
from the top of this mountain, parallel to the earth's axis,
will meet the superficies of the sea, (supposing that superficies continued inland from the Mediterranean,) must be
about 2382 toises in horizontal distance, or about 2$\frac{1}{2}$ minutes south of the summit, that is, in the parallel of 45⁰ 46$\frac{1}{2}$′ ;
and if this parallel be continued all round the globe, the
points of the earth's surface between it and the equator, are
all more distant from the earth's axis than the top of Mont
Blanc is ; whereas all the points to the north of it are nearer
to that axis. A stone, therefore, from the top of Mont Blanc,

natural history of the earth. " Rivers," says this author, " do not carry into the sea the spoils which they bring from the land, but employ them in the formation of deltas of low alluvial land at their mouths, according to what Major Rennell has proved." The fact of the formation of *deltas* from the spoils which the rivers carry from the higher grounds, is perfectly ascertained ; and the detail into which Major Rennell has entered in the passage referred to by Mr Kirwan, does credit to the acuteness and accuracy of that excellent geographer. But it is not there asserted, that rivers employ *all* the materials which they carry with them, in the formation of those deltas, and deliver none of them into the sea. On the contrary, they carry from the *delta* itself mud and earth, which they can deposit nowhere but in the sea ; and it is this circumstance chiefly that limits the increase of those allu-

---

if carried any where to the south of the above parallel, will retard the earth's diurnal motion ; but if carried any where to the north of the same line, will accelerate that motion.

The same quantity of matter, however, carried an equal distance toward the pole, and toward the equator, from any point, will lose more velocity in the former case than it will gain in the latter, as easily follows from the nature of the circle. Therefore, supposing an equal dispersion of the detritus of a mountain in all directions, the parts that go toward the pole will most disturb the diurnal motion ; and hence a balance on their side, or in favour of acceleration, as already observed.

vial lands, and makes them either cease to increase,
or makes them increase very slowly after a certain
period, though the supply of earth from the higher
grounds remains nearly the same.   To make Mr
Kirwan's argument conclusive, it would be necessary to prove, that *all* the mud carried down by
the Nile or the Ganges, was deposited on the low
lands before these rivers enter the sea ; a thing so
obviously absurd, that nothing but his haste to obtain a conclusion unfavourable to the Plutonic system, could have prevented him from perceiving
it. *

379. A remark which Major Rennell has made
concerning the mouths of rivers, in his Geography
of Herodotus, deserves Mr Kirwan's attention,
though perhaps he may not be able to put on it an
interpretation quite so favourable to his system.
The remark is, that the mouths of great rivers are
often formed on principles quite opposite to one
another, so that some of them have a real delta or
triangle of flat land at their mouths, while others

---

* The instance mentioned in the Geological Essays, from
the travels of the Abbé Fortis, concerning urns thrown into
the Adriatic, upwards of 1400 years ago, and not yet covered with mud, must be explained from peculiar circumstances, or local causes, with which we are unacquainted, as it
makes against the deposition of earth near the shore, and in
narrow seas ; a general fact, which, I think, every body admits.

have an estuary, or what may not improperly be called a *negative* delta. Of the latter kind are some of the greatest rivers in the world, the Plata, the Oronoco and the Maranon, and by far the greatest number of our European rivers. Nobody can doubt, that the three rivers just named carry with them as much earth as the Nile, or the Euphrates, or any other river in the world. All this they have deposited in the sea, and committed to the currents, which sweep along the shore of the American continent, and by these they have been spread out over the unlimited tracts of the ocean.

Indeed, nothing can be more just than Dr Hutton's observation, that where low land is formed at the mouths of rivers, there the rivers bring down more than the sea is able to carry away ; but that where such land is not formed, it is because the sea is able to carry off immediately all the deposit which it receives.

880. Mr Kirwan has denied on another principle the power of the sea to carry to a distance the materials delivered into it : " Notwithstanding," says he, " many particles of earth are by rivers conducted to the sea, yet *none are conveyed to any distance*, but are either deposited at their mouths, or rejected by currents or by tides ; and the reason is, because the tide of flood is always more impetuous and forcible than the tide of ebb, the advancing waves being pressed forward by the countless num-

ber behind them, whereas the retreating are pressed backward by a far smaller number, as must be
evident to an attentive spectator ; and hence it is
that all floating things cast into the sea, are at last
thrown on shore, and not conveyed into the mid
regions of the sea, as they should be if the recipro-
cal undulations of the tides were equally powerful." *

381. But if the *attentive spectator*, instead of
trusting to a vague impression, or listening to some
crude theory of undulations, reflects on one of the
most simple facts respecting the ebbing and flowing
of the tides, he will be very little disposed to ac-
quiesce in the above conclusion.   He has only to
consider, that the flowing of the tide requires just
six hours, and the ebbing of it likewise six hours ;
so that the same body of water flows in upon the
shore, and retreats from it, in the same time.   The
quantity of matter moved, therefore, and the velocity with which it is moved, are in both cases the
same ; and it remains for Mr Kirwan to show in
what the difference of their force can possibly consist.

The force with which the waves usually break
upon our shores, does not arise from the velocity of
the tide being greater in one direction than in
another.   In the main ocean, the waves have no

---

* Kirwan's Geol. Essays, p. 439.

4

progressive motion, and the columns of water alternately rise and fall, without any other than a reciprocating motion : a kind of equilibrium takes
place among the undulations, and each wave being
equally acted upon by those on opposite sides, remains fixed in its place.   Near the shore this cannot happen ; the water on the land side from its
shallowness being incapable of rising to the height
necessary to balance the great undulations which
are without.   The water runs, therefore, as it
were, from a higher to a lower level, spreading itself towards the land side.   This produces the
breakers on our shores, and the surf of the tropical
seas.   A rock or a sandbank coming within a certain distance of the surface, is sufficient, in any
part of the ocean, to obstruct the natural succession
of undulations ; and, by destroying the mutual reaction of the waves, to give them a progressive instead of a reciprocating motion.

382. It is, however, but from a small distance,
that the waves are impelled against the shore with
a progressive motion.   The border of breakers that
surrounds any coast is narrow, compared with the
distance to which the *detritus* from the land is
confessedly carried ; the water, while it advances
at the surface, flows back at the bottom ; and these
contrary motions are so nearly equal, that it is but
a very momentary accumulation of the water that is
ever produced on any shore.

If it were otherwise, and if it were true that the sea throws out every thing, and carries away nothing, we should have a constant accumulation of earth and sand along all shores whatsoever, at least wherever a stream ran into the sea. This, as is abundantly evident, is quite contrary to the fact.

So, also, the bars formed at the mouths of rivers, after having attained a certain magnitude, increase no farther, not because they cease to receive augmentations from the land, but because their diminution from the sea, increasing with their magnitude, becomes at length so great, as completely to balance those augmentations. When properly examined, therefore, the phenomena, which have been proposed as most inconsistent with the indefinite transportation of stony bodies, afford very satisfactory proofs of that operation.

383. It is true, that bodies which float in the water, when carried along on the tops of the waves towards a shelving beach, having acquired a certain velocity, are thrown farther in upon the land than the distance they would have floated to, if they had been simply sustained by the water. The depth of water, therefore, at the place where they take the ground, is not likely to be such as to float them again, and to carry them out towards the sea. They are, therefore, left behind ; and this produces an appearance of a force impelling floating bodies to-

wards the land, much greater and more general than really takes place.

These observations may serve to show, how unsound the principles are from which Mr Kirwan's conclusions are deduced : they are perhaps more than is necessary for that purpose : it might have been sufficient to observe, that the increase of land on the sea shore is limited, though the augmentation from the land is certainly indefinite, a proof that the diminution from the sea is constant and equal to the increase.

384. "Mariners," says Mr Kirwan, "were accustomed, for some centuries back, to discover their situation, by the kind of earth or sand brought up by their sounding plummets ; a method which would prove fallacious, if the surface of the bottom did not continue invariably the same." *

The fact here stated, that mariners, when navigation was more imperfect than it is now, had very frequent recourse to this method, and that they still use it occasionally, is very true. But from this, the only inference that can be fairly deduced is, that the changes at the bottom of the sea are very slow, and the variation but little ; not merely from one year to another, but even from one century to another. The rules by which the mariner judged of his position from the quality of the earth

---

* Geol. Essays, p. 440.

which the lead brought up, and which were deduced no doubt from observations made at no very
great distance of time, might be sufficient for his
purpose, though a slow change had been all the
while going forward.   Such observations could at
best have little accuracy, and could not be affected
by small variations.   It is the slowness of the
change, that makes the experience of one age applicable, in this, as in innumerable other instances,
to the observations of the next.   If a long interval
is taken, we will look in vain for the same uniformity of results.   A pilot, who would at present
judge of his position in the German Ocean, by
comparing his soundings with those taken by
Pytheas (supposing them known) in his navigation
of that sea, more than 2000 years ago, could hardly be expected to determine his latitude and longitude with great exactness ; and I know not if the
most zealous advocate for the immutability of the
earth's surface, would be willing to trust his safety in a ship that was guided by such antiquated
rules.

NOTE XX. § 118.

Inequalities in the Planetary Motions.

385. The assertion that, in the planetary motions, we discover no mark, either of the com-

mencement or termination of the present order, refers to the late discoveries of Lagrange and Laplace, which have contributed so much to the perfection of physical astronomy. From the principle
of universal gravitation, these mathematicians have
demonstrated, that all the variations in our system
are periodical ; that they are confined within certain limits ; and consist of alternate diminution
and increase. The orbits of the planets change
not only their position, but even their magnitude
and their form : the longer axis of each has a slow
angular motion ; and, though its length remains
fixed, the shorter axis increases and diminishes, so
that the form of the orbit approaches to that of a
circle, and recedes from it by turns. In the same
manner, the obliquity of the ecliptic, and the inclination of the planetary orbits, are subject to change;
but the changes are small, and, being first in one
direction, and then in the opposite, they can never
accumulate so as to produce a permanent or a progressive alteration. Thus, in the celestial motions,
no room is left for the introduction of disorder ; no
irregularity or disturbance, arising from the mutual action of the planets, is permitted to increase
beyond certain limits, but each of them, in time,
affords a correction for itself. The general order
is constant, in the midst of the variation of the
parts ; and, in the language of Laplace, there is a
certain mean condition, about which our system

perpetually *oscillates*, performing small vibrations
on each side of it, and never receding from it far.*
The system is thus endowed with a stability, which
can resist the lapse of unlimited duration ; it can
only perish by an external cause, and by the introduction of laws, of which at present no vestige is
to be traced.

386. The same *calculus* to which we are indebted for these sublime conclusions, informs us of two
circumstances, which mark the law here treated of
as an effect of wise design, to the entire exclusion
both of necessity and chance.   One of these circumstances consists in the planetary motions being
all in the same direction, or all *in consequentia*, as
it is called by the astronomers.   This is essential to
the compensation and stability above mentioned : †
had one planet circulated round the sun in a direction from east to west, and another in a direction from west to east, the disturbances they would
have produced on one another's motion would not
necessarily have been periodical ; their irregularities might have continually increased, and they
might have deviated in the course of ages from
their original condition, beyond any limits that can
be assigned.

---

* Exposition du Systeme du Monde, par Laplace, Livre
IV. Chap. VI. p. 199, 2d edit.
    † Laplace, ibid.

The other circumstance, on which the stability of our system depends, is the small eccentricity of the planetary orbits, or their near approach to circles. Were their orbits very eccentric, an opening would be given to progressive change, that might so far increase, as to prove the destruction of the whole. But neither the movement of all the planets in the same direction, nor the small eccentricity of their orbits, can be ascribed to accident, since that either of these should happen by chance, in as many instances as there are planets, both primary and secondary, is almost infinitely improbable. Again, that any necessity in the nature of things should have either determined the *direction* of the planetary motions, or proportioned the *quantity* of them to the intensity of the central force, cannot be admitted, as these are things unavoidably conceived to be quite independent of one another. It remains, therefore, that we consider the laws, which make the disturbances in our system correct themselves, and by that means give firmness and permanence to it, as a proof of the consummate wisdom with which the whole is constructed.

387. The geological system of Dr Hutton, resembles, in many respects, that which appears to preside over the heavenly motions. In both, we perceive continual vicissitude and change, but confined within certain limits, and never departing far

from a certain mean condition, which is such, that, in the lapse of time, the deviations from it on the one side, must become just equal to the deviations from it on the other.   In both, a provision is made for duration of unlimited extent, and the lapse of time has no effect to wear out or destroy a machine, constructed with so much wisdom.   Where the movements are all so perfect, their beginning and end must be alike invisible.

NOTE XXI. § 122.

Changes in the apparent Level of the Sea.

388. In speaking of the natural epochas marked out by the phenomena of the mineral kingdom, we have supposed a greater simplicity, and separation of effects from one another, than probably takes place in nature.   We have, for instance, abstracted, in speaking of the waste and degradation of the land, from that elevation which may have been carried on at the same time.   This appeared necessary to be done, in order to simplify as much as possible the view that was to be given of the whole ; but there can be no doubt, that, while the land has been gradually worn down by the operations on its surface, it has been raised up by the expansive forces acting from below.   There is even reason

to think, that the elevation has not been uniform, but has been subject to a kind of oscillation, insomuch, that the continents have both ascended and descended, or have had their level alternately raised and depressed, independently of all action at the surface, and this within a period comparatively of no great extent.

It will be easily understood, that the facts we are going to state, each taken singly, prove nothing more than a change of the line in which the surface of the sea intersects the surface of the land, leaving it uncertain to which of the two the change ought really to be ascribed. Taken in combination, however, these facts may determine what each of them separately cannot ascertain. I shall first, therefore, mention some of the principal observations relative to the change above mentioned, and shall then compare them, in order to discover whether it is most probable that this change has been produced by the motion of the land or of the sea.

389. If we begin with examining the coasts of our own island, we shall find clear evidence every where, that the sea once reached higher up upon the land than it does at present. The marks of an ancient sea beach are to be seen beyond the present limits of the tide, and beds of sea shells, not mineralized, are found in the loose earth or soil, sometimes as high as thirty feet above the pre-

sent level of the sea. Some of these on the shores
of the Frith of Forth are very well known, and
have been often mentioned. Indeed, on the shores
of that frith, many monuments appear, which would
seem to carry the difference between the present
and the ancient level of the sea, to more than forty feet. The ground on which the Botanic Gar-
den of Edinburgh is situated, after a thin covering
of soil is removed, consists entirely of sea sand,
very regularly stratified, with layers of a black carbonaceous matter, in thin lamellæ, interposed between them. Shells I believe are but rarely
found in it, but it has every other appearance of a
sea beach. The height of this ground above the
present level of the sea is certainly not less than
forty feet.

390. On almost every part of the coast where
the rocks do not rise quite abrupt and precipitous
from the sea, similar marks of the lowering of the
sea, or the rising of the land, may be observed.
On the shores opposite to ours, the same appearances are remarked. The author of the Lettre
Critique to M. de Buffon, tells us, that he had
found the bottom of a bason at Dunkirk, which
he had reason to think was dug about 950 years
ago, ten feet and a half above the present low
water mark, though it must have been originally
under it. The bottom of this bason is in the native chalk. From this, the same author concludes,

that the sea at Dunkirk lowers its level at the rate
of an inch nearly in seven years. The observation was made in 1762, (Lettre à M. le Comte de
Buffon, &c. p. 55.) *

391. The shores of the Low Countries, and of
Holland, have been often instanced in proof of the
same kind of changes, and it has been supposed,
that, independently of those artificial barriers
which at present exclude the waters of the ocean
from overflowing a great part of this tract, nature
herself has brought it nearer to the surface than
it had formerly been. It is indeed certain, that
those countries, to a very great extent inland, have
either been under the sea at some period, by no
means remote if compared with the great revolutions of the globe, or that they are entirely alluvial, and of the same sort with the Deltas formed
at the mouths of rivers. The relative changes,
however, of the sea and land on this tract, have
been differently represented, and I am unwilling,
on that account, to found any argument on them.

---

* In the county of Suffolk, near Wood Bridge, at the dis-
stance of seven or eight miles from the sea, are the Crag-
pits, in which prodigious quantities of sea shells are discovered, many of them perfect and quite solid, (Pennant's Arctic Zoology, Introd. p. 6.)   Lincolnshire affords various
proofs of the same kind ; but some other circumstances in
the appearance of that coast, just about to be taken notice of,
indicate changes of a more complicated nature.

392. If we proceed farther to the north, to the shores of the Baltic for instance, we have undoubted evidence of a change of level in the same direction as on our own shores. The level of this sea has been represented as lowering at so great a rate as 40 inches in a century. Celsius observed, that several rocks which are now above water, were not long ago sunken rocks, and dangerous to navigators; and he particularly took notice of one, which, in the year 1680, was on the surface of the water, and in the year 1731 was $20\frac{1}{2}$ Swedish inches above it. From an inscription near Aspô, in the lake Melar, which communicates with the Baltic, engraved, as is supposed, about five centuries ago, the level of the sea appears to have sunk in that time no less than 13 Swedish feet. * All these facts, with many more which it is unnecessary to enumerate, make the gradual depression, not only of the Baltic, but of the whole northern ocean, a matter of certainty.

393. Supposing these changes of level between the sea and land to be sufficiently ascertained, the supposition which at first occurs is, that the motion has been in the sea rather than in the land, and that the former has actually descended to a lower level. The imagination naturally feels less difficulty in conceiving, that an unstable fluid like the

---

* Frisii Opera, Tom. III. p. 274.

sea, which changes its level twice every day, has undergone a permanent depression in its surface, than that the land, the *terra firma* itself, has admitted of an equal elevation.   In all this, however, we are guided much more by fancy than reason ; for, in order to depress or elevate the absolute level of the sea, by a given quantity, in any one place, we must depress or elevate it by the same quantity over the whole surface of the earth ; whereas no such necessity exists with respect to the elevation or depression of the land.   To make the sea subside 30 feet all round the coast of Great Britain, it is necessary to displace a body of water 30 feet deep over the whole surface of the ocean.   The quantity of matter to be moved in that way is incomparably greater than if the land itself were to be elevated ; for though it is nearly three times less in specific gravity, it is as much greater in bulk, as the surface of the ocean is greater than that of this island.

394. Besides, the sea cannot change its level, without a proportional change in the solid bottom on which it rests.   Though there be reason to suppose that such changes in the bottom do actually take place, yet they are probably much slower and more imperceptible than those which we are here considering.   It is evident, therefore, that the simplest hypothesis for explaining those changes of level, is, that they proceed from the motion, up-

wards or downwards, of the land itself, and not from that of the sea. As no elevation or depression of the sea can take place, but over the whole, its level cannot be affected by local causes, and is probably as little subject to variation as any thing to be met with on the surface of the globe.

395. Other observations, however, made on different shores from the preceding, give greater certainty to this conclusion, and make it clear, that the motion or change which we are now treating of is not to be ascribed to the sea itself.

The observations just mentioned prove, that the level of the North Sea is lower now than it was heretofore ; but it appears, that in the Mediterranean, the opposite takes place. Very accurate observations made by Manfredi, render it certain, that the superficies of the Hadriatic was higher about the middle of the last century, than toward the beginning of the Christian era.

Some repairs that were carrying on in the cathedral church of Ravenna, in the year 1731, afforded him an opportunity of observing, that the ancient, and probably original, pavement, was four feet and a half below the present, and nearly a foot under the level of the sea at high water. * Now, when the church was built, this cannot have been

---

* Commentarii Academiæ Bononiensis, Tom. II. pars 1ma, p. 237, &c. and pars 2da, p. 1, &c.

the position of the pavement, relatively to the level of the sea, for it would have subjected the floor to be under water twice in twenty-four hours, and must have done so the more unavoidably, because at that time (the beginning of the fifth century) the walls of Ravenna were washed by the sea. The fact that this pavement is under the high-water mark, by the quantity just mentioned, was ascertained by actual levelling. This result was confirmed by similar facts, observed by Zendrini at Venice.

396. Manfredi himself attributes all this to the elevation of the surface of the sea, and has entered into a long calculation to ascertain at what rate that surface may be supposed to rise, on account of the earth and sand brought down by the rivers, and spread out over the bottom of the sea. But as the fact of the rise of the level of the sea is not general, and as the contrary is observed in the north seas, as already proved, this hypothesis will not explain the apparent rise in the level of the Hadriatic.

397. Though a local subsidence, or settling of the ground, could hardly account for this change, the pavement being perfect in its level, and the walls of the cathedral without any shake, yet a subsidence that has extended to a great tract, as to the whole of Italy, if the mass moved has continued parallel to itself, and changed its place slow-

ly, will agree very well with the appearances. The facts here stated are also the more deserving of attention, that about Ravenna, the land, at the same time that it has sunk in its level, has extended its surface, and has encroached on the sea. Since the time of Augustus, the line of the coast has been carried farther out by about three miles. * This last is the undoubted effect of the degradation of the land by the rivers ; and here we have very clear evidence of the forces, both under and above the surface, producing their respective effects at the same time, so that while the surface is raised by earth brought down by the rivers, every given point in the ground is depressed and let down to a lower level. †

398. On the southern coast of Italy similar facts have been observed. Breislac, in his Topografia Fisica della Campagna di Roma, ‡ from certain appearances in the Gulfs of Baja and Naples, concludes, that at the beginning of the Christian æra, the level of the sea was lower on that part of the coast than it is now. The facts which he mentions are the following : 1mo, The remains of an

---

* Manfredi, *ibid.*

† On the coast of Dalmatia also, the rising of the level of the sea has been remarked, particularly at the ruins of Diocletian's palace of Spalatro.

‡ Cap. vi. p. 300.

ancient road are now to be seen in the Gulf of
Baja at a considerable distance from the land.
2do, Some ancient buildings belonging to Porto
Giulio are at present covered by the sea. 3tio,
Ten columns of granite at the foot of Monte
Nuovo, which appear to have belonged to the Temple of the Nymphs, are also nearly covered by the
sea. 4to, The pavement of the Temple of Serapis is now somewhat lower than the high water
mark, though it cannot be supposed that this edifice when built was exposed to the inconvenience
of having its floor frequently under water. 5to,
The ruins of a palace, built by Tiberius in the
island of Caprea, are now entirely covered by the
sea.

Thus, it appears that the level of the sea is sinking
in the more northern latitudes, and rising in the Mediterranean, and it is evident that this cannot happen by the motion of the sea itself. The parts of
the ocean all communicating with one another,
cannot rise in one place and fall in another; but, in
order to maintain a level surface, must rise equally
or fall equally over the whole of its extent. If,
therefore, we place any confidence in the preceding observations, and they are certainly liable to no
objection, either from their own nature or the character of the observers, we must consider it as demonstrated, that the relative change of level has
proceeded from the elevation or depression of the

land itself. This agrees well with the preceding
theory, which holds, that our continents are subject
to be acted upon by the expansive forces of the
mineral regions; that by these forces they have been
actually raised up, and are sustained by them in
their present situation.

399. According to some other facts stated by
the same ingenious author, it appears, that on the
coast of Italy the progress of the sea in ascending,
or of the land in descending, has not been uniform
during the period above mentioned, but that different oscillations have taken place; so that, from
about the beginning of the Christian era, till some
time in the middle ages, the sea rose to be sixteen
feet higher than at present, from which height it
has descended till it became lower than it is now,
and from that state of depression it is now rising
again. Breislac infers this from two facts, which
he combines very ingeniously with the preceding,
viz. the remains of some ancient buildings, at the
foot of Monte Nuovo, five or six feet above the
present level of the sea, in which are found the
shells of some of those little marine animals that
eat into stone : And again, the marble columns of
the temple of Serapis, which are also perforated by
pholades, to the height of sixteen feet above the
ground. All these changes Breislac ascribes to the
motion of the sea itself; a supposition which, as we
have seen, cannot possibly be admitted, since no-

thing can permanently affect the level of the sea in one place, which does not affect it in all places whatsoever.

400. Appearances, which indicate such alternations as have just been mentioned in the level of the sea, are to be met with on some other coasts. In England, on the coast of Lincolnshire, the remains of a forest have been observed, which are now entirely covered by the sea. * The submarine stratum which contains the remains of this forest, can be traced into the country to a great distance, and is found throughout all the fens of Lincolnshire. The stratum itself is about four feet thick ; it is covered in some places by a bed of clay sixteen feet thick, and under it for twenty feet more is a bed of soft mud, like the scourings of a ditch, mixed with shells and silt.

Here then we have a stratum which must have been once uppermost on the surface of the dry land, though one part of it is now immersed under the sea, and another covered with earth, to the depth of sixteen feet. A change of level in the sea itself will not explain these appearances : they can only be explained by supposing the whole tract of land to have subsided, which is the hypothesis adopted by the author of the description in the Transactions, M. Corria de Serra ; the subsidence,

* Phil. Trans. 1799, p. 145.

however, is not here understood to arise from the
mere yielding of some of the strata immediately
underneath, but is conceived to be a part of that
geological system of alternate depression and elevation of the surface, which probably extends to the
whole mineral kingdom. To reconcile all the different facts, I should be tempted to think, that the
forest which once covered Lincolnshire, was immersed under the sea by the subsidence of the land
to a great depth, and at a period considerably remote ; that when so immersed, it was covered over
with the bed of clay which now lies on it, by deposition from the sea, and the washing down of earth
from the land ; that it has emerged from this great
depth till a part of it has become dry land ; but that
it is now sinking again, if the tradition of the country deserves any credit, that the part of it in the sea
is deeper under water at present than it was a few
years ago. This might also serve to reconcile, in
some measure, the phenomena of this submarine
forest with the appearances which indicate an extension of the land on the coast of Lincolnshire.
Indeed the extension of the land is no direct proof,
either of its own elevation, or of the depression of
the sea, as we may conclude from the instance of
Ravenna already mentioned.

401. We have concluded from the facts stated
above, that the level of the sea rises in the Mediterranean, and sinks in the more northern lati-

1

tudes ; and thence some have suspected, that the
level of the sea had in general a tendency to rise
towards the equator, and to sink towards the poles.
This is the notion of Frisi, as has been already remarked, and he suggests, that this rise of the sea
may be owing to a slight acceleration in the earth's
diurnal motion.   But there are facts which show,
that between the tropics the relative level of the
sea and land has sunk, and is lower at present than
it was at some former period, probably not extremely remote.   The opinion of Frisi, therefore,
is unsupported by observation, and, as has been already shown, cannot be justified from theory.

Between the tropics, islands are formed from the
mere accumulation of coral ; and it is the peculiarity of those regions, to produce rocks that have
not passed through the usual process of mineral
consolidation. *   The islots, however, which are
thus formed, must have their bases laid on a solid
rock, though perhaps at a great depth ; and it is
not probable, that after they are once raised above
the surface of the sea, they can still rise farther,
except by some elevation of the rock which serves

* Dr Foster, in his Voyage round the World, (Vol. II. p.
146,) gives an instance in the South Sea Islands, where the
surface of the island, though entirely a coral rock, was raised forty feet above the level of the sea.

as their foundation. *   Now, at Palmerston island, which comprehends nine or ten low islots, that may be reckoned the heads of a great reef of coral rock, Captain Cook informs us of his having seen, " far beyond the reach of the sea, even in the most violent storms, elevated coral rocks, which, on examination, appeared to have been perforated in the same manner that the rocks are that now compose the outer edge of the reef. This evidently shows," he adds, " that the sea had formerly reached so far ; and some of these perforated rocks were almost in the centre of the island." †

The same excellent navigator, giving an account of the peninsula at Cape Denbigh, remarks : " It appeared to me, that this peninsula must have been an island in remote times ; for there were marks of the sea having flowed over the isthmus."

402. We are here touching on one of those subjects, where we feel much the want of accurate and ancient observations, and where it is not from the infancy, but the maturity of science that any thing approaching to certainty can be looked for.   The utmost that we can expect at present, is an antici-

---

* A very curious account of the formation of such islands is given by A. Dalrymple, Esq. in the Philosophical Transactions, Vol. LVII. p. 394.
† Cook's Third Voyage, Vol. I. p. 221.

pation, which future ages must certainly modify and correct. The best thing, in the mean time, that can be done for the advancement of this branch of geological knowledge, is to ascertain with exactness the relative level of the sea, and of such points upon the land as can be distinctly marked, and pointed out to succeeding ages. This is not so easy as it may at first appear. Where every object changes, it is difficult to find a measure of change, or a fixed point from which the computation may begin. The astronomers already feel this inconvenience, and when they would refer their observations to an immoveable plane, that shall preserve its position the same in all ages, they meet with difficulties, which cannot be removed but by a profound mathematical investigation.

In geology, we cannot hope to be delivered from this embarrassment in the same manner; and we have no resource but to multiply observations of the difference of level; to make them as exact as possible, and to select points of comparison that have a chance of being long distinguished. The improvements in barometrical measurements, which give such facility to the determination of heights, along with so considerable a degree of accuracy, will furnish an accumulation of facts that must one day be of great value to the geologist.

NOTE XXII. § 123.

Fossil Bones.

403. The remains of organized bodies, at present included in the solid parts of the globe, may be divided into three classes. The first consists of the shells, corals, and even bodies of fish, and amphibious animals, which are now converted into stone, and make integrant parts of the solid rock. All these are parts of animals that existed *before the formation of the present land*, or even of the rocks whereof it consists. These remains have been already treated of, and the evidence which they furnish must ever be regarded as of the utmost importance in the theory of the earth. The second class consists of remains, which, by the help of stalactitical concretions, are converted into stone. These are the *exuviæ* of animals, which existed on the very same continents on which we now dwell, and are no doubt the most ancient among their inhabitants, of which any monument is preserved. In comparison of the first class, they must, nevertheless, be considered as of very modern origin.

404. The third class consists of the bones of animals found in the loose earth or soil ; these have not acquired a stony character, and their na-

ture appears to be but little changed, except by the progress of decomposition and of mouldering into earth. No decided line can be drawn between the antiquity of this and the preceding class, as there may be between the preceding and the first. In some instances, the objects of this third class may be coëval with those of the second; in general, they must be accounted of later origin, as they are certainly not preserved in a manner so well fitted for long continuance.

405. The animal remains of the second class, are generally found in the neighbourhood of limestone strata, and are either enveloped or penetrated by calcareous, or sometimes ferruginous matter. Of this sort are the bones found in the rock of Gibraltar, and on the coast of Dalmatia. The latter are peculiarly marked for their number, and the extent of the country over which they are scattered, leaving it doubtful whether they are the work of successive ages, or of some sudden catastrophe that has assembled in one place, and overwhelmed with immediate destruction, a vast multitude of the inhabitants of the globe. These remains are found in greatest abundance in the islands of Cherso and Osero; and always in what the Abbé Fortis calls an *ocreostalactitic earth*. The bones are often in the state of mere splinters, the broken and confused relics of various animals, concreted with fragments of marble

and lime, in clefts and chasms of the strata.*
Sometimes human bones are said to be found in
these confused masses.

406. A very remarkable collection of bones in
this state is found in the caves of Bayreuth in Fran-
conia. Many of these belong, as is inferred with
great certainty from the structure of their teeth, to
a carnivorous animal of vast size, and having very
little affinity to any of those that are now known.
The bones are found in different states, some being
without any stalactitical concretion, and having the
calcareous earth still united to the phosphoric acid,
so that they belong to the third, rather than the second, of the preceding divisions. In others, the
phosphoric acid has wholly disappeared, and given
place to the carbonic.

The number of these bones, accumulated in the
same place, is matter of astonishment, when it is
considered, that the animals to which they belonged were carnivorous, so that more than two can
never have lived in the same cavern at the same
time. The caves of Bayreuth seem to have been
the den and the tomb of a whole dynasty of unknown monsters, that issued from this central spot
to devour the feebler inhabitants of the woods,
during a long succession of ages, before man had

* Travels into Dalmatia, p. 449.

subdued the earth, and freed it from all domination but his own.

407. The fossil bones of the second and third class, but chiefly of the third, have now afforded matter of conjecture and discussion for more than a century. The facts with respect to them are very numerous and interesting, but can be considered here only very generally.

The remains of this kind, consist of the bones only of large animals, so that they have generally been compared with those of the elephant, the rhinoceros, the hippopotamus, or other animals of great size. The bones of smaller animals have also been found, but much more rarely than the other. It is usually remarked, that the bones thus discovered in the earth are larger than those of the similar living animals.

Another general fact concerning these remains, is, that they are found in all countries whatsoever, but always in the loose or travelled earth, and never in the genuine strata. Since the year 1696, when the attention of the curious was called to this subject, by the skeleton of an elephant dug up in Thuringia, and described by Tentzelius, * there is hardly a country in Europe which has not afforded instances of the same kind. Fossil bones, particularly grinders and tusks of elephants, have been

---

* Phil. Trans. Vol. XIX. p. 757.

found in other places of Germany, in Poland, France, Italy, Britain, Ireland, and even Iceland. * Two countries, however, afford them in greater abundance by far than any other part of the known world; namely, the plains of Siberia in the old continent, and the flat grounds on the banks of the Ohio in the new. †

408. When the bones in Siberia were first discovered, they were supposed to belong to an animal that lived under ground, to which they gave the name of the *mammouth;* and the credit bestowed on this absurd fiction, is a proof of the strong desire which all men feel of reconciling extraordinary appearances with the regular course of nature. Much skill, however, in natural history was not required to discover that many of the bones in question resembled those of the elephant, particularly the grinders and the tusks of that animal. Others resembled the bones of the rhinoceros; and a head of that kind, having the hide preserved upon it, was found in Siberia, and is still in the imperial cabinet at Petersburgh.

Pallas has described the fossil bones which he found in the museum at Petersburgh, on his being appointed to the superintendence of it, and enu-

---

* A grinder of an elephant found in Iceland, is described by Bartholinus, Acta Hafniens. Vol. I. p. 83.

† The fossil bones on the Ohio are described in two papers by Mr P. Collinson, Phil. Trans. Vol. LVII. p. 464 and 468.

merates, not only bones that belong, in his opinion,
to the elephant and rhinoceros, but others that belong to a kind of buffalo, very different from any
now known, and of a size vastly greater. * He has
also described, in another very curious memoir, the
bones of the same kind that he met with in his travels through the north-east parts of Asia.

The fossil bones found on the banks of the Ohio,
resemble in many things those of Siberia; like them
they are contained in the soil or alluvial earth, and
never in the solid strata; like them too they are no
otherwise changed from their natural state, than by
being sometimes slightly calcined at the surface;
they are also of great size, and in great numbers,
being probably the remains of several different species.

409. Two inquiries concerning these bones have
excited the curiosity of naturalists; first, to discover among the living tribes at present inhabiting
the earth, those to which the fossil remains may
with the greatest probability be referred; and, secondly, to find out the cause why these remains
exist in such quantities, in countries where the
animals to which they belong, whatever they be,
are at present unknown. The solution of the first
of these questions, is much more within our reach

---

* Novi Comment. Petrop. Tom. XIII. (1768,) p. 436, and
Tom. XVII. p. 576, &c.

than the second, and at any rate must be first sought for.

On the authority of so eminent a naturalist as Pallas, the bones from Siberia may safely be referred to the elephant, the rhinoceros, and buffalo, as mentioned above, though perhaps to varieties of them with which we are not now acquainted. With respect to the bones of North America, the question is more doubtful, for they have this particular circumstance attending them, viz. that along with the thighbones, tusks, &c. which might be supposed to belong to the elephant, grinders are always found of a structure and form entirely different from the grinders of that animal. * Some naturalists, particularly M. Daubenton, referred these grinders to the hippopotamus ; but Dr W. Hunter appears to have proved, in a very satisfactory manner, that they cannot have belonged to either of the animals just mentioned, but to a *carnivorous* animal of enormous size, the race of which, fortunately for the present inhabitants of the earth, seems now to be entirely extinct. † The foundation of Dr Hunter's opinion is, that in these grinders the enamel is merely an external covering ; whereas, in the elephant, and other animals des-

---

* See Mr Collinson's papers, above referred to. Phil. Trans. Vol. LVII.

† Phil. Trans. Vol. LVIII. p. 3, &c.

tined to live on vegetable food, the enamel is intermixed with the substance of the tooth. *

410. Though this argument appears to be of
considerable weight, yet Camper, who was greatly
skilled in comparative anatomy, and who had studied this subject with particular attention, was of
opinion, that these grinders belong to a species of
elephant. This opinion he states in a letter to
Pallas, who had found grinders and other bones of
this same animal, on the western declivity of the
Ural mountains. † Camper denies that the animal is carnivorous, because the *incisores*, or canine
teeth, are wanting; and he argues farther, from
the weight of the head, which may be inferred
from the weight of the grinders, that the neck
must have been short, and the animal must have
been furnished with a *proboscis*. He afterwards
abandoned the latter hypothesis, and gave it as his

---

* A fossil grinder, in the collection of John Macgowan,
Esq. of Edinburgh, answers nearly to Mr Collinson's description, and is very well represented by the figure which
accompanies it. This grinder weighs four pounds one-fourth
avoirdupois; the circumference of the *corona* is eighteen
inches; the coat of enamel is one-fourth of an inch thick;
there are five double teeth; in Mr Collinson's specimen
there are only four.

† Acta Acad. Petrop. Tom. I. (1777,) pars posterior,
p. 213, &c.

opinion, that the *incognitum* was neither carnivorous, nor a species of the elephant. *

411. Nevertheless, Cuvier, in a *mémoire* read
before the National Institute of Paris, maintains,
that the fossil bones of the new Continent, as well
as most of those of the old, belong to certain species of the elephant ; of which, at least, two do not
now exist, and are only known from remains preserved in the ground. He distinguishes them
thus : †

> *Elephas mammonteus,—maxillâ obtusiore, la-
> mellis molarium tenuibus, rectis.*
> *Elephas Americanus,—molaribus multicuspidi-
> bus, lamellis post detritionem quadri-lobatis.*

The latter species, which is meant to include
the *animal incognitum*, is said to have lived, not
only in America, but in many parts of the old
Continent. Yet some late inquiries into the structure of the teeth of graminivorous animals, and
particularly of the elephant, make it very improbable that the *incognitum* has belonged to this
genus. ‡ The grinders of the elephant have been

Ibid. Tom. II. (1784,) p. 262.

† Mémoires de l'Institut National, Sciences Physiques,
Tom. II. p. 19, &c.

‡ See Mr Home's Observations on the Teeth of Graminivorous Animals, Phil. Trans. 1799. Also, an Essay on the
Structure of the Teeth, by Dr Blake.

found to consist of three substances, enamel, bone, and what is called the *crusta petrosa*, applied in layers, or folds contiguous to one another ; and no vestige of this structure appears in the grinders of the unknown animal of the Ohio. * At the same time, Dr Hunter's assertion, that this animal was carnivorous, is rendered doubtful, not only by the want of *canine* teeth, but also from the resemblance between its grinders and those of the wild boar, which Mr Home has observed to be considerable. † The grinder of the boar is similar to that of the elephant, in the extent of the masticating surface, but not at all in the internal structure ; and the same is true of the tooth of the *animal in-*

---

* In a paper inserted in the fourth volume of the American Philosophical Transactions, an account is given of two different grinders that are found at the Salt Licks near the Ohio. One of them resembles the grinder of the elephant, and may have belonged to the Elephas Americanus of Cuvier ; the other agrees pretty nearly with the grinder of Dr Hunter's *animal incognitum.* The author of the paper thinks that the *animal incognitum* was not wholly carnivorous, as the *incisores*, or canine teeth, are never found. At the Great Bone Lick, bones of smaller animals, particularly of the buffalo kind, have been discovered. The saline impregnation of the earth at these Licks must no doubt have contributed to the preservation of the bones. Trans. American Phil. Soc. Vol. IV. (1799,) p. 510, &c.

† Observations on the Grinding Teeth of the Wild Boar and *Animal Incognitum.* Phil. Trans. 1801, p. 319.

cognitum, so that a considerable probability is established, that it and the boar are of the same genus, and both destined to live occasionally either on animal or vegetable food.

412. Another *animal incognitum* found in South America has been described by Cuvier, and appears to be of a different genus from the *incognitum* of the North. Thus, if we include the two *incognita* of America, the *elephas mammonteus,* the unknown buffalo of Pallas, and the great animal of Bayreuth, we have at least five distinct genera, or species of the animal kingdom, which existed on our continents formerly, but do not exist on them now. The number is probably much greater : Pallas mentions fossil horns of a gazelle, of an unknown species ; and horns of *deer* are often found, that cannot be referred to any species now existing. Those extinct races have been remarkable for their size : some of the ancient elephants appear to have been three times as large as any of the present. *

413. The inhabitants of the globe, then, like all the other parts of it, are subject to change. It is not only the individual that perishes, but whole *species,* and even perhaps *genera,* are extinguished. It is not unnatural to consider some part of this change as the operation of man. The exten-

* Camper, Nov. Acta Petrop. Tom. II. (1784,) p. 257.

sion of his power would necessary subvert the balance that had before been established between the
inhabitants of the earth, and the means of their
subsistence. Some of the larger and fiercer animals might indeed dispute with him, for a long
time, the empire of the globe; and it may have required the arm of a Hercules to subdue the monsters which lurked in the caves of Bayreuth, or
roamed on the banks of the Ohio. But these, with
others of the same character, were at length exter-
minated : the more innocent species fled to a distance from man ; and being forced to retire into
the most inaccessible parts, where their food was
scanty, and their migration checked, they may
have degenerated from the size and strength of
their ancestors, and some species may have been
entirely extinguished.

But besides this, a change in the animal kingdom seems to be a part of the order of nature, and
is visible in instances to which human power cannot have extended. If we look to the most ancient inhabitants of the globe, of which the remains are preserved in the strata themselves, we
find in the shells and corals of a former world
hardly any that resemble exactly those which exist
in the present. The species, except in a few instances, are the same, but subject to great varieties.
The vegetable impressions on slate, and other argillaceous stones, can seldom be exactly recog-

nised ; and even the insects included in amber, are
different from those of the countries in which the
amber is found.

414. Supposing, then, the changes which have
taken place in the qualities and habits of the animal creation, to be as great as those in their structure and external form, we can have no reason to
wonder if it should appear that some have formerly dwelt in countries from which the similar races
are now entirely banished.   The power of living in
a different climate, of enduring greater degrees of
cold or of heat, or of subsisting on different kinds of
food, may very well have accompanied the other
changes.   Though one species of elephant may
now be confined to the southern parts of Asia,
another may have been able to endure the severer
climates of the north ; and the same may be true
of the buffalo or the rhinoceros.   In all this no
physical impossibility is involved ; though whether
it is a probable solution of the difficulty concerning the origin of these animal remains, can only be
judged of from other circumstances.

415. If we consider attentively the facts that
respect the Siberian fossil bones, there will appear
insurmountable objections to every theory that
supposes them to be exotic, and to have been
brought into their present situation from a distant
country.

The extent of the tract through which these

bones are scattered, is a circumstance truly wonderful.   Pallas assures us, * that there is not a river of considerable size in all the north of Asia,
from the Tanais, which runs into the Black Sea, to
the Anadyr, which falls into the Gulf of Kamt-
chatka, in the sides or bottom of which bones of
elephants and other large animals have not been
found.   This is especially the case where the rivers run in plains through gravel, sand, clay, &c. ;
among the mountains, the bones are rarely discovered.   The extent of the tract just mentioned
exceeds four thousand miles ; and how the bones
could be distributed over all that extent, by any
means but by the animals having lived there, it
seems impossible to conceive.   No torrent nor
inundation could have produced this effect, nor
could the bones brought in that way have been
laid together so as to form complete skeletons.

416. One fact recorded by the same author,
seems calculated to remove all uncertainty.   It is
that of the carcase of a rhinoceros, almost entire,
and covered with the hide, found in the earth in
the banks of the river Wilui, which falls into the
Lena below Jakutsk. †   Some of the muscles and

---

* De Reliquiis Animalium exoticorum, per Asiam Bo-
realem repertis.—Nov. Comment. Petrop. Tom. XVII.
(1772,) p. 576.

† Pallas *ubi supra*, p. 586.   Also, Voyages de Pallas,
Tom. IV. p. 131.

tendons were actually adhering to the head when Pallas received it.   The head, after being dried in an oven, is still preserved in the museum at Petersburgh.   The preservation of the skin and muscles of this natural mummy, as Pallas calls it, was no doubt brought about by its being buried in earth that was in a state of perpetual congelation; for the place is in the parallel of 64°, where the ground is never thawed but to a very small depth below the surface.

But by what means can we account for the carcase of a rhinoceros being buried in the earth, on the confines of the polar circle ?   Shall we ascribe it to some immense torrent, which, sweeping across the desarts of Tartary, and the mountains of Altai, transported the productions of India to the plains of Siberia, and interred in the mud of the Lena the animals that had fed on the banks of the Barampooter or the Ganges ?   Were all other objections to so extraordinary a supposition removed, the preservation of the hide and muscles of a dead animal, and the adhesion of the parts, while it was dragged for 2000 miles over some of the highest and most rugged mountains in the world, is too absurd to be for a moment admitted.   Or shall we suppose that this carcase has been floated in by an inundation of the sea, from some tropical country now swallowed up, and of which the numerous islands of the Indian Archipelago are the remains ?

The heat of a tropical climate, and the putrescence naturally arising from it, would soon, independently of all other accidents, have stripped the bones of their covering. Indeed this *instantia singularis*, as in every sense it may properly be called, seems calculated for the express purpose of excluding every hypothesis but one from being employed to explain the origin of fossil bones. It not only excludes the two which have just been mentioned, but it excludes also that of Buffon, viz. that these bones are the remains of animals which lived in Siberia, when the arctic regions enjoyed a fine climate, and a temperature like that which southern Asia now possesses. From the preservation of the flesh and hide of this rhinoceros, it is plain, that when the body was buried in the earth, the climate was much the same that it is now, and the cold sufficient to resist the progress of putrefaction.

Pallas takes notice of the inconsistency of the state of this skeleton, with the hypothesis of Buffon; but he does not observe that the inconsistency is equally great between it and his own hypothesis, the importation of the fossil bones by an inundation of the sea, and that flesh or muscle must have been entirely consumed long before it could be carried by the waves to the parallel of 64°, from any climate which the rhinoceros at present inhabits.

417. The presence of petrified marine objects in

places where some of the fossil bones are found, is
no proof that the latter have come from the sea,
though it is produced as such both by Pallas himself, and afterwards by Kirwan. These marine
bodies are the shells and corals that have been
parts of calcareous rocks, from which being detached by the ordinary progress of disintegration,
they are now contained in the beds of sand or gravel where the animal remains are buried. They
have nothing in common with these remains; they
are real stones, and belong to another, and a far
more remote epocha. Such objects being found in
the same place where the bones lie, argues only
that the strata in the higher grounds, from which
the gravel has come, are calcareous; and nothing
can show in a stronger light the necessity of distinguishing the different condition of fossil bodies,
united by the mere circumstance of contiguity, before we draw any inference as to their having a
common origin. If the marine remains were in
the same condition with the bones; if they were in
no respect mineralized; then the conclusion, that
both had been imported by the sea, would have
great probability; but without that, their present
union must be held as casual, and can give no in-
sight into the origin of either.

418. On the whole, therefore, no conclusion remains, but that these bones have belonged to species of elephants, rhinoceros, &c. which inhabited

the very countries where their remains are now buried, and which could endure the severity of the Siberian climate. The rhinoceros of the Wilui certainly lived on the confines of the Polar Circle, and was exposed to the same cold while alive, by which, when dead, its body has been so long, and so curiously preserved.

These animals may also have lived occasionally farther to the south, among the valleys between the great ranges of mountains that bound Siberia on that side. Fossil bones are but rarely found in these valleys, probably because they have been washed down from thence into the plains. We must observe, too, that those animals may have migrated with the seasons, and by that means avoided the rigorous winter of the high latitudes. The dominion of man, by rendering such migration to the larger animals difficult or impossible, must have greatly changed the economy of all those tribes, and narrowed the circle of their enjoyments and existence. The heaps in which the fossil bones appear to be accumulated in particular places, especially in North America, have a great appearance of being connected with the migrations of animals, and the accidents that might bring multitudes of them into the same spot.

What holds of Siberia and of North America, is applicable, *a fortiori*, to all the other places where animal remains are found in the same condition.

Thus we are carried back to a time when many larger species of animals, now entirely extinct, inhabited the earth, and when varieties of those that are at present confined to particular situations, were, either by the liberty of migration, or by their natural constitution, accommodated to all the diversities of climate. This period, though beyond the limits of ordinary chronology, is posterior to the great revolutions on the earth's surface, and the latest among geological epochas.

NOTE XXIII. § 128.

Geology of KIRWAN *and* DELUC.

419. The two champions of the Neptunian system, who have distinguished themselves most by their hostility to Dr Hutton, are Deluc and Kirwan. They have carried on their attack nearly on the same plan, and have employed against their antagonist the weapons both of theology and science. With a spirit as injurious to the dignity of religion, as to the freedom of philosophical inquiry, they have disregarded a maxim enforced by the authority of Bacon, and by all our experience of the past ; " *Tanto magis hæc vanitas inhibenda venit et coërcenda, quia, ex divinorum et humanorum male-sana admixtione, non solum educitur philo-*

*sophia phantastica, sed etiam religio hæretica.
Itaque salutare admodum est, si mente sobriâ,
fidei tantum dentur quæ fidei sunt."* *

Proceeding, accordingly, in direct opposition to
rules that have never yet been violated with impunity, and mistaking the true object of a theory of
the earth, they carry back their inquiries to a period
prior to the present series of causes and effects,
where, having neither experience nor analogy to
direct them, they pretend to be guided by a superior light. They would have us to consider their
geological speculations as a commentary on the text
of Moses ; they endeavour to explain the action of
creative power, and, with indiscreet curiosity, would
tear off the veil which the hand of the prophet has
so wisely respected. But the veil cannot be torn
off, and all that is behind it must be to man as that
which never has existed.

420. M. Deluc has nevertheless treated very dif-
fusely of the history of the solar system, previous

---

* The whole passage is deserving of attention, and it
seems as if the prophetic spirit of Bacon had addressed it to
the cosmologists of the present day. " *Pessima enim res est
errorem* APOTHEOSIS, *et pro peste intellectûs habenda est, si
vanis accedat veneratio. Huic autem vanitati nonnulli ex mo-
dernis summâ levitate ita indulserunt, ut, in primo capitolo*
GENESEOS, *et aliis Scripturis Sacris, philosophiam naturalem
fundare conati sunt :* Inter VIVA quærentes MORTUA."—Nov.
Organum, Lib. I. Aphor. 65.

to the establishment of the present laws of nature, and has dwelt on it with great complacency, and singular minuteness of detail. His tenth letter to Lametherie has the following title :

" On the History of the Earth, from the time when that planet was penetrated by *light*, till the appearance of the sun ; a portion of time which includes the origin of heat, and of the figure of the earth ; of its primeval strata, of the ancient sea, of our continents, as the bottom of that sea, of the great chains of mountains, and of vegetation." *

I must confess that I am unacquainted with every thing of this letter but the title ; and could not easily be prevailed on to follow any man who professedly goes out of nature in search of knowledge ; who pretends to give the history of our planetary system when there was no sun, and to enumerate the events which took place between the existence of that luminary, and the existence of light. The absurdity of such an undertaking admits of no apo-

---

* Journal de Physique, Tom. XXXVII. (1790,) Partie 2de, p. 332. As I may not have done justice to this extraordinary title, it may be right to present it in the original. " Sur l'Histoire de la TERRE, depuis que cette planette fut penetrée de LUMIERE, jusqu'à l'apparition du SOLEIL ; espace de tems qui renferme les ORIGINES de la *chaleur,* et de la *figure* de notre globe ; de ses *couches primordiales,* de *l'ancienne mer,* de nos *continens,* comme fond de cette mer, de leurs grandes chaînes de *montagnes,* et de la *vegetation.*"

logy ; and the smile which it might excite, if addressed merely to the fancy, gives place to indignation when it assumes the air of philosophic investigation.

421. It sets, however, in a strong light, the inconsistencies that may be observed in the intellectual character of the same individual, to consider
that the author of this strange and inconsistent re-
verie, is, nevertheless, an excellent observer, and
well skilled in experimental inquiries. It will
hardly be believed that he who writes the history
of the earth before the formation of the sun, is
versed in the principles of inductive reasoning ;
and that he has added much to the stock of geological knowledge, having observed accurately, and
described with great perspicuity and candour. His
Lettres Physiques are full of valuable and just observations, though accompanied with reasonings
that do not seem always entitled to the same praise ;
and in another work he has succeeded where many
men of genius had failed, and has made considerable improvements in a branch of the mathematics,
without borrowing almost any assistance from the
principles of that science. *

422. Some of the same observations apply to Mr
Kirwan. His Geological Essays have also for their
object to explain the first origin of things ; and to

---

* Essai sur les Modifications de l'Atmosphère.

say that he has not succeeded, in an attempt where no man ever can succeed, implies no reproach on the execution of his work, whatever it may do on the design. We have indeed no criterion by which the execution of it can be estimated : what would in any other place be a blemish, may be here deserving of praise ; and if the work is full of confusion and perplexity, these are qualities inherent in the subject which it is intended to describe. It were, no doubt, to be wished, that after emerging into the regions of day, Mr Kirwan had been as successful in copying the beauty and simplicity of nature, as in representing the disorder and inconsistency of the chaotic mass. But his cosmology is without unity in its principles, or consistency in its parts : the causes introduced, are, for the most part, such as will account for one set of appearances just as well as for another ; or, if any of them is likely to prove inadequate to the effect ascribed to it, a new and arbitrary hypothesis is always ready to come to its assistance. The information given is seldom exact : a multitude of facts brought together, without the order and discussion essential to precise knowledge ; and an infinity of quotations, amassed without criticism or comparison, afford proofs of extensive reading, but of the most hasty and superficial inquiry. Thus we have seen passages from Ulloa and Frisi, produced in support of opinions, which, when fairly stated, they had the most direct tendency to overthrow.

423. In one respect, the geological writings of Kirwan are far inferior to Deluc's : They are evidently the productions of a man who has not seen nature with his own eyes ; who has studied mineralogy in cabinets, or in books only ; but who has seldom beheld fossils in their native place. With the balance in his hand, and the external characters of Werner in his view, he has examined minerals with diligence, and has discovered many of those marks which serve to ascertain their places, in a system of artificial arrangement. But to *reason* and to *arrange* are very different occupations of the mind ; and a man may deserve praise as a mineralogist, who is but ill qualified for the researches of geology.

424. The same hurry and impatience are visible in the manner in which his argument against Dr Hutton is usually conducted. He has seldom been careful to make himself master of the opinions of his adversary ; and what he gives as such, and directs his reasonings against, have often no resemblance to them whatsoever. Without any intention to deceive others, but deceived himself, he usually begins with misrepresenting Dr Hutton's notions, and then proceeds to the refutation of them. In this imaginary contest, it will readily be supposed, that he is in general successful : when a

man has the framing both of his own argument, and that of his antagonist, he must be a very unskilful logician if he does not come off with the advantage.

425. It is but justice, however, to the Neptunists, to acknowledge, that they are not all liable to the censure of beginning their researches from a period antecedent to the existence of the laws of nature. This absurdity does not, so far as I know, infect the system of Werner. That mineralogist has not proposed to explain the first origin of things, though he has supposed, at some former period, a condition of the globe very unlike the present, viz. the entire submersion of the solid under the fluid part.

NOTE XXIV. § 129.

System of BUFFON.

426. The affinity of Dr Hutton's theory to that of Buffon, is nothing more than what arises from their making use of the same agents, viz. fire and water, in producing the present condition of the earth's surface. In almost all other respects the two theories are extremely different. The order in which those agents are employed in them, is directly opposite, as has already been remarked ; Buffon

introducing the action of fire first, and of water only in the second place, to waste and destroy mineral bodies, and afterwards to dispose them anew, and arrange them into strata. He makes no provision for the consolidation of these strata, nor any for their angular elevation; he has no means of explaining the unstratified rocks; nor any, but one extremely imperfect, for explaining the inequalities of the earth's surface.

Again, Buffon mistook, in some degree, the true object of a theory of the earth; and though he did not go back, like the geologists just named, to a time when the laws of nature were not fully established, he begins from a condition of things too unlike the present to be the basis of any rational speculation. He does not, indeed, undertake to examine the state of our planetary system before the sun existed; for from such extravagance, even when most disposed to indulge his fancy, he would surely have revolted. But he treats of the world, when the earth and the planets had just ceased to be a part of the sun, and were newly detached from the body of that luminary. *

This hypothesis concerning the origin of the planets, contrived chiefly to account for the circumstance of their motion being all in the same direc-

---

* According to Buffon, the granite is the true solar matter, unchanged but by its congelation.

tion, and in other respects not only unsupported,
but even inconsistent with the principle of gravitation, has nothing in common with a theory, confined as Dr Hutton's is, within the field which
must for ever bound our inquiries, and not venturing to speculate about the earth, when in a condition totally different from the present.

427. In what relates to the future, the two systems are not more like than in what relates to the
past.   Buffon represents the cooling of our planet,
and its loss of heat, as a process continually advancing, and which has no limit, but the final extinc-
tion of life and motion over all the surface, and
through all the interior, of the earth.   The death
of nature herself is the distant but gloomy object
that terminates our view, and reminds us of the
wild fictions of the Scandinavian mythology, according to which, *annihilation* is at last to extend
its empire even to the gods.   This dismal and
unphilosophic vision was unworthy of the genius
of Buffon, and wonderfully ill suited to the ele-
gance and extent of his understanding.   It forms
a complete contrast to the theory of Dr Hutton,
where nothing is to be seen beyond the continuation of the present order ; where no latent seed of
evil threatens final destruction to the whole ; and
where the movements are so perfect, that they
can never terminate of themselves.   This is surely
a view of the world more suited to the dignity of

NATURE, and the wisdom of its AUTHOR, than has yet been offered by any other system of cosmology.

428. I have often quoted Buffon in the course of these Illustrations, and most commonly for the purpose of combating his opinions; but I am very sensible, nevertheless, of the obligations under which he has laid all the sciences connected with the natural history of the earth.

The extent and variety of his knowledge, the justness of his reasonings, the greatness of his views, his correct taste, and manly eloquence, qualified him, better, perhaps, than any other individual, to compose the History of Nature. The errors into which he has fallen, are almost all the unavoidable consequences of the circumstances in which he was placed; and if their amount is estimated by the proportion that they bear to the general excellence of the work, they will be reckoned but of small account. Buffon began to write when many parts of natural history had made but little progress; when the quantity of authentic information was small, and when scientific and correct description was hardly to be found. Many of the greatest and most important facts in geology were quite unknown, and scarcely any part of the mineral kingdom had been accurately surveyed; and, with such materials as this state of things afforded, it is not wonderful if some parts of the edifice he erected have not proved so solid and durable as the rest.   Had he appeared somewhat later ; had he been farther removed from the time when reasonings *a priori* usurped the place of induction ; and had he been as willing to correct the errors into which he had been betrayed by imperfect information, as he was ingenious in defending them, his work would probably have reached as great perfection, as it is given for any thing without the sphere of the accurate sciences to attain.   If he had examined the natural history of the earth more with his own eyes, and been as careful to delineate it with fidelity as force ; if he had listened with greater care to the philosophers around him ; had he attended to the demonstrations of Newton more, and despised the arrangements of Linnæus less ; he would have produced a work, as singular for its truth as for its beauty, and would have gone near to merit the eulogy pronounced by the enthusiasm of his countrymen, MAJESTATI NATURÆ PAR INGENIUM.

NOTE XXV. § 130.

Figure of the Earth.

429.   That the earth is a spheroidal body, compressed at the poles, or elevated at the equator, is

a fact established by many accurate experiments; and though these experiments do not exactly co-incide, as to the degree of oblateness which they give to that spheroid, they agree sufficiently to put it beyond all dispute, that the earth, though solid, has nearly the same figure which it would assume if fluid, in consequence of its rotation on its axis.

Now, it is not at all obvious, to what physical cause this phenomenon is to be ascribed. The earth, as it exists at present, has none of the conditions that render the assumption of the figure of equilibrium in any way necessary to it. Constituted as it is, its parts cohere with forces incomparably too great to obey the laws of statical pressure, or to assume any one figure rather than another, on account of the centrifugal tendency which results from its revolution on its axis. There is no necessity that its superficies should be every where level, or perpendicular to the direction of gravity, nor that every two columns, standing on the same base, any where within it, and reaching from thence to any two points of the surface, should be of such weights as precisely to balance one another. Neither of these, indeed, is at all conformable to fact. They are, however, the very suppositions on which the determination of the spheroid of equilibrium is founded; and as they certainly do in no degree belong to the earth, it seems strange that the result deduced from them

should be in any way applicable to it. This coincidence remains, therefore, to be explained;
and it must greatly enhance the merit of any geological system, if it can connect this great and
enigmatical phenomenon with the other facts in
the natural history of the earth.

430. To establish such a connection, has, accordingly, been a favourite object with geologists, whether they have embraced the Neptunian or Volcanic
theory : both have thought that they were entitled
to suppose the primeval fluidity of the globe, the
one by water, and the other by fire ; and in whatsoever way that fluidity was produced, the result of
it could be no other than the spheroidal figure of
the whole mass, agreeably to the laws of hydrostatics.
If in this fluid state the earth was homogeneous, the
spheroid would be accurately elliptical, and the compression at the poles would be $\frac{1}{230}$ of the radius of
the equator; if the fluid was denser toward the
centre, the flattening would be less : and in either
case, the body, as it acquired solidity, may be supposed to have retained its spheroidal figure with
little variation.   But though the fluidity of the
earth will account for the phenomenon of its oblate
figure, it may reasonably be questioned, whether
this fluidity can be admitted, in consistency with
other appearances.   According to what is established above, none of the appearances in the mineral

I

kingdom indicate more than a partial fluidity in any former condition of the earth. The present strata, made up as they are of the ruins of former strata, though softened by heat, have not been rendered fluid by it, and have even possessed their softness in parts, and in succession, not altogether, nor at the same time.

The unstratified, and more crystallized substances, were cast in the bosom of others, which were solid at the time when they were fluid. In all this, therefore, there is no indication of a fluidity prevailing through the whole mass, or even over the whole surface of the earth, and therefore nothing that can explain the spheroidal figure which it has acquired. The supposition, then, of the entire body of the earth, or even of its external crust, having been fluid, though it might account for the compression at the poles, does not connect that fact with the other facts in the natural history of the globe, and fails, therefore, in the point most essential to a theory. It is liable, also, to other objections, whether it be conceived to have proceeded from fire or from water ; whether it has happened on the principles of Buffon or of Werner.

431. First, let us suppose that the fluidity of the earth, or of the external crust of it, at least to a certain depth, proceeded from a solution of the whole in the waters of the ocean ; and, waving all the objections that have been stated to this hypothesis, on

account of the absolute insolubility of many mineral substances in water, let us suppose them all soluble in a certain degree, and let us compute the quantity of the menstruum, which, on the suppositions most favourable to the system, must have been required to this great geologico-chemical operation.

The siliceous earth, though not soluble in water *per se*, yet, after being dissolved in that fluid by means of an alkali, was found by Dr Black, in his analysis of the Geyser water, to remain suspended in a quantity of water, between 500 and 1000 times its own weight. This is one of the facts most favourable to the Neptunian theory ; and that every advantage may be given to that theory, we shall take the least of the numbers just mentioned, and suppose that siliceous earth may be dissolved or suspended in 500 times its weight of water.

Taking this for the extreme degree of insolubility of mineral substances, (though there are many of which the insolubility is absolute, or, to speak in the language of calculation, infinitely great,) we may suppose the insolubility of all the rest, or the quantities of water in which they are dissolved, to be ranged in a descending scale from 500 to 0, the extreme degree of deliquescence. Then, taking the arithmetical mean between these extremes, it will give us 250, as the proportion of water in which mineral substances may at an average be dissolved. But this average is much less than the truth ; for the quantity of siliceous earth is great in comparison of any of the rest, and the mineral substances that are extremely soluble in water are but in a small quantity; therefore, when we suppose mineral bodies, at a medium, to be soluble in 250 times their own weight of water, we make a supposition extremely favourable to the Neptunian system.

432. This is the proportion between the *weight* of the solvent, and of the substances held in solution : to have the proportion of their *bulks,* we may suppose the specific gravity of mineral bodies in general to be to that of water as 5 to 2, and then we have the ratio of bulks, that of 250×5 to 2×1, or of 625 to 1. It follows, then, that minerals in general cannot be supposed soluble in less than 625 times their bulk of water.

433. Again, it must be allowed to the Neptunists, that the fluidity of the whole earth is not necessary to account for its assuming the spheroidal figure. It is sufficient if the whole of that crust or shell of matter was fluid, which is contained between the actual surface of the terrestrial spheroid, and the surface of the sphere inscribed within it; that is, of the sphere which has for its diameter the polar axis of the earth. The whole of the minerals which compose this shell, must at least have been dissolved in water, and have formed the chaotic mass of Mr Kirwan. The volume

of the water required for this was not less than 625 times the bulk of the spheroidal shell that has just been mentioned.

But, assuming the difference between the polar axis and the equatorial diameter to be $\frac{1}{300}$ of the latter, which is the supposition most agreeable to the phenomena, it is easy to show that the magnitude of the above spheroidal shell, or the difference between the solid content of the earth, and the sphere inscribed in it, is greater than $\frac{1}{151}$, and less than $\frac{1}{150}$ of the whole earth; so that the earth is less than 151 times the spheroidal shell.

The volume of the water, therefore, necessary to hold in solution the materials of this shell, is to the volume of the whole earth as 625 to 151, or in a greater ratio than that of four to one : and such, therefore, at the very least, is the quantity of water which Mr Kirwan supposes, after it ceased to act in its chemical capacity, to have retired into caverns in the interior of the earth. Thus the Neptunists, in their account of the spheroidal figure of the earth, are reduced to a cruel dilemma, and are forced to choose between a physical and a mathematical impossibility.

If we would inquire whether the opinion of the igneous origin of minerals, as commonly received by the Vulcanists, is capable of affording a better

8

solution of this difficulty, the theory of M. de Buffon is the first that presents itself.

434. That philosopher considers the existence of the spheroidal figure as a proof that the whole of the earth must have been originally fluid ; and as the fluidity of the whole can only be ascribed to fusion, he has supposed that the earth was originally a mass of melted matter struck off from the sun by the collision of a comet ; and that this mass, when made to revolve on its axis, put on a spheroidal figure, which it has retained, though now cooled down to congelation.

This system need not be considered in detail ; the foundation of it is laid in such defiance of the principles of geometry and mechanics, that the architect, notwithstanding all the fertility of his invention, and all the resources of his genius, was never able to give any solidity to the structure.

But it will be said, that we may take a part of the system, without venturing on the whole, and may suppose that the earth, or at least the external crust of it, has been fluid by fire, though we do not inquire into the cause of this fire, or into the manner in which it was produced.

It is indeed true, that, when this is done, we have not the same sort of absurdity to encounter that we met with in the Neptunian system, and that the Vulcanic theory does not, like it, come into direct collision with an axiom of geometry.

There are, nevertheless, great objections to it; for though all the phenomena of the mineral kingdom attest a fluidity of igneous origin, yet it is a fluidity that was never more than partial; and though it has been over all the earth, has been over it in succession only. Besides, we are not entitled to assume the existence, and again the disappearance of such a great quantity of heat, without assigning some cause for the change.

435. Since, then, neither the hypothesis of the Neptunists or the Vulcanists, affords any good explanation of the figure of the earth, or such a one as can connect it with the other appearances in its natural history, it remains to inquire, whether the system that supposes a partial and successive fluidity, like Dr Hutton's, has any resource for explaining this great phenomenon.

Of this subject Dr Hutton has not treated; and when I was first made acquainted with his system, it appeared to me a very serious objection to it, that it did not profess to give an explanation of so important a fact as the oblate figure of the earth. On considering the matter more closely, however, I found that there were principles contained in it from which a very satisfactory solution (and, I think, the only satisfactory solution) of that difficulty might be deduced. This solution I shall endeavour to explain, in as far, at least, as is necessary for the purpose of general illustration.

It is laid down in Dr Hutton's theory, that the surface of the earth is perpetually changed by the *detritus* of the land ; and that from the materials thus afforded, new horizontal strata are perpetually formed at the bottom of the sea. If this be true, and if the alternations of decay and renovation have been often repeated, it is certain, that the figure of the earth, whatever it may have originally been, must be brought at length to coincide with the spheroid of equilibrium.

436. Here it is necessary to remark, that the expressions, *figure of the earth*, and *surface of the earth*, are each of them occasionally taken in two different senses.

The surface of the earth, in its most obvious sense, is that which bounds the whole earth, and includes all its inequalities ; it is a surface extremely irregular, rising to the tops of the mountains, descending to the bottoms of the valleys, and having the continuity of its curvature often interrupted, or suddenly changed. This may be called the *actual* surface, and the figure bounded by it, the *actual* figure, of the earth.

The surface of the earth, in another sense, is one that is every where horizontal, and is the same which water assumes when at rest.

This superficies is determined by the circumstance of its being constantly perpendicular to the direction of gravity ; it is the surface marked out

by levelling, and may be supposed to be continued
from the sea, through the interior of the land, till
it meet the sea again.   The figure bounded by this
horizontal surface, may properly be called the *statical* figure of the earth.

When it is said that the figure of the earth is an
oblate spheroid, it is the statical, not the actual figure which is meant ; and the degrees of the meridian which astronomers measure, are also referred
to the superficies of the former.

437. Suppose now a body like the earth, but
with its actual figure infinitely more irregular, having a sea circumfused around it, the water will descend into the lowest situations, and will so arrange itself, that its surface shall be perpendicular
every where to the plumb-line, or to the direction of
gravity, in which state only it can remain at rest.
The figure of the superficies which the sea must
thus take will be of a continuous curvature, and
will return into itself; though it may, if the actual figure is very irregular, be far either from a
sphere or a spheroid.   If, however, we suppose the
solid parts of this mass subject to be dissolved or
worn away, and carried down to the ocean, there
will be a tendency to give to the whole body the
same figure that it would have assumed, if it had
been entirely fluid, and subject to the laws of hydrostatics.   This tendency is the result of two
principles.

438. Let us suppose the body just described to have no rotation, so that the particles of it are actuated only by the forces of cohesion and of attraction.

It is then clear, that every particle taken away by attrition from the parts above the level of the sea, and deposited under the surface of it, makes the general figure more compact, bringing the remoter parts nearer to the centre of gravity of the whole; so that, in time, if the body is homogeneous, all the points of the surface will become equally distant from that centre. Thus the *actual* figure changes continually, and approaches nearer to the *statical*.

While this change is going forward in the actual figure, there is another produced on the statical, that tends very much to accelerate the final coincidence of the two.

The effect of the inequalities of the land, that rise above the horizontal surface, is, by their attraction, to render the parts of that surface immediately under them, more convex, *cæteris paribus*, than the rest. Again, where there are parts of extraordinary depth in the sea, that is, where the solid and denser parts are far removed from the surface of the ocean, the curvature of the superficies of the sea is thereby diminished, and that superficies is rendered less convex than it would be if the sea were shallower. These propositions are both capable of strict mathematical demonstration. Hence the taking away of any particle of matter from the top of a mountain tends to diminish the curvature of the horizontal surface under the mountain, where it is greatest ; and the deposition of the same particle at the bottom of the sea, tends to increase the curvature of this superficies where it is least.   The general tendency, therefore, being to increase the curvature where it is least, and to diminish it where it is greatest, must be to bring about an uniform curvature throughout, that is, a spherical figure.   Thus, by the waste and subsequent stratification of the land, the direction of gravity is continually altered ; it is more and more concentrated, and the figure brought nearer to that which a fluid would assume.

439. If now we suppose the body to revolve on its axis, all other things remaining as before, the surface bounding the sea will become different from what it was in the former case, and will be more swelled out toward the middle or equatorial regions.   The land above the level of the sea will still, as before, be worn down and deposited in the bottom of the sea, so as to form strata nearly parallel to its surface : the tendency, therefore, is to render the real figure of the planet nearer to the statical.   At the same time the *statical* figure is changed, as explained above ; so that the two figures mutually approach, and the limit, or ulti-

mate figure to which they tend, is one over which
the ocean might be diffused every where to the
same depth, for then the causes of change would
entirely cease. But this figure is no other than
the spheroid of equilibrium, which, therefore, is
the effect which the waste and reconsolidation of
the land would necessarily produce, if the process
were continued indefinitely, without interruption.
In this, as in many other instances, when a body is
subject to the action of causes by which its form is
gradually changed, the figure best adapted to resist those changes, is the figure which the changes
themselves ultimately produce.

Also, whatever be the irregularities of density,
the tendency to a change of figure will not cease
till the body is moulded into that particular spheroid which admits of being covered with water
every where to the same depth. * Thus it appears,

---

* In the same manner as a transition is thus made from
an irregular figure to a spheroid of equilibrium, so, if the
actual figure were at first more simple than the spheroid, it
would still be changed into this last by degrees.

Let us conceive, for instance, that the earth is at rest, and
is a perfect sphere of solid matter, surrounded by an ocean
every where of equal depth, for example, of one mile. Then,
if a rotatory motion be communicated to it, so that it shall
revolve on its axis in twenty-four hours, in consequence of
the centrifugal force, the water circumfused about the sphere
will immediately rise up under the equator, and will be-

that a solid of an irregular figure, and of irregular
density, provided it be in part covered with water;

---

come part of a spheroidal surface, (not elliptical, but nearly
so,) the equatorial diameter of which is greater than the polar axis, in the ratio of 588 to 577.   By this means the water will be accumulated at the equator to the depth of nearly
2.5 miles, and form a zone surrounding the earth, and extending about 37° on each side of the equator.   The remainder of the surface will be left dry, forming two vast
circumpolar continents, that reach 53° on every side of the
poles, and that are elevated in the middle more than four
miles above the level of the sea.

Such would be the state of our globe, on the hypothesis
above laid down ; and, if there were no waste or destruction of the land, this order of things would be permanent,
and neither the solid nor fluid part of the mass could ever
acquire any other figure than that which has been described.
But, if the same laws be supposed to regulate the action of
the atmosphere in those circumstances, that do actually regulate it according to the present constitution of the globe,
the vapours raised up from the surface of the sea, would be
carried by the winds over the land, where they would be
condensed and precipitated in rain.   Thus, all the agents
of destruction would be let loose on the two great circumpolar continents ; rivers would be formed ; the land would become deeply intersected by ravines ; those ravines would
gradually open into wide valleys ; the masses of greatest resistance would be shaped into hills and mountains : and
from a superficies originally smooth and uniform, the same
inequalities would be produced which at present diversify
the surface of the earth.

While the parts of the sphere without the spheroid are

and be at the same time subject to waste above the surface of the sea, and reconsolidation under it, has a tendency to acquire, in time, the same figure that it would have acquired had it been entirely fluid.

440. In the preceding reasonings, we have supposed the process of decay and subsequent stratification to be carried on without interruption, till the whole of the land is covered by the sea. This supposition is useful for explaining the nature of the forces which have determined the figure of the earth; but there is no reason to think that it has ever been realized in its full extent, the elevation of strata from the bottom of the sea interrupting the progress, and producing new land in one place, as the old decays in another. The very same land also, which is wasted at its surface, may perhaps be lifted up by the forces that are placed under it;

---

thus continually diminished, the loose earth and sand washed down from them, will be deposited at the bottom of the sea, and will form strata parallel to the surface of the superincumbent water. The actual and statical figure are thus brought nearer one another; and, at the same time the statical is changed, on the principle already explained, (the change in the direction of gravity,) and is made continually to approximate to a state, which when it has attained, no farther change can take place, viz. an oblate elliptic spheroid, of which the surface is perpendicular to the direction of gravity, having the equatorial diameter to the polar axis in the ratio of 230 to 229.

or it may be let down, undergoing alterations of
its level, from causes that we do not perceive, but
of which the action is undoubted, (§ 388.) But
notwithstanding these interruptions, the general
tendency to produce in the earth a spheroidal figure
may remain, and more may be done by every revolution, to bring about the attainment of that
figure than to cause a deviation from it. This
figure, therefore, though never likely to be perfectly acquired, will be the *limiting* or *asymptotic*
figure, if it may be so called, to which the earth
will continually approach.

441. If the preceding conclusions are just, and
if the figure of equilibrium is only an asymptotic
figure, to which that of the earth may approximate,
but cannot perfectly attain, we are not to be surprised if considerable deviations from it are actually observed. This has accordingly happened, insomuch, that the results deduced from the most
accurate measurement of degrees of the meridian,
differ from one another, in the oblateness they give
to the earth, by nearly one half of the quantity to
be determined. When we compare the degrees
measured in France, and in some other countries
of Europe, with those measured in Peru, we obtain for the compression at the poles, less than
$\frac{1}{300}$ of the radius of the earth. But when we compare the degrees measured in France with one

another, and with those lately measured in England, we find that they are best represented by a pheroid that has its compression $\frac{1}{150}$ of its semiaxis. *   There is reason to think, therefore, that the meridians are not elliptical ; and other observations seem to show, that they are not even similar to one another ; or that the earth is not, strictly speaking, a solid of revolution ; so, also, the comparison of the degree measured at the Cape of Good Hope, with those measured on the opposite side of the equator, creates a suspicion, that the northern and southern hemispheres are not perfectly alike, and that the earth is not equally compressed at the Arctic and the Antarctic poles. These irregularities, though they do not affect the general fact of the earth's compression at the poles, show that the true statical figure is but imperfectly attained ; and though this may be accounted for, without having recourse to the principles involved in our theory, it is in a manner very unsatisfactory, and, by help of suppositions, not at all consistent with the original fluidity ascribed to the whole mass, or to the exterior crust of the earth.

442. As the principles here laid down explain how a solid body may attain very nearly the figure

---

* Exposition du Système du Monde, par Laplace, p. 61, 2d edit.

which a fluid would acquire in order to preserve
its parts in equilibrio ; and since the oblate figure
belongs to other of the planets as well as the earth,
and the globular to all the great bodies of the universe, this suggests an analogy that goes deep into
the economy of nature, and extends far beyond the
limits within which the mineralogist is wont to
confine his speculations.

443. That no very irregular figure is found
among the planetary bodies, may therefore be considered as a proof of the universality of that system
of waste and reconsolidation that we have been endeavouring to trace in the natural history of the
earth. A farther proof of the same arises from
considering, that for every given mass of matter,
having a given period of rotation, there are two
different spheroids that answer the conditions of
establishing an equilibrium among its parts, the
one near to the sphere, and the other very distant
from it, and so oblate, as to have a lenticular form.
Thus the earth, supposing it homogeneous, might
either be in equilibrio, by means of the figure
which it actually has, or of one in which the polar
was to the equatorial diameter as 1 to 768. The
same is true of the other planets ; and yet we no
where find that this highly compressed spheroid is
actually employed by nature. The reason, no
doubt, is, that in so oblate a spheroid, the equilibrium between the gravitating and the centrifugal

force is of the kind that does not re-establish itself when disturbed ; so that the parts let loose, and not kept in their place by firm cohesion, would fly off altogether. In such a body, the waste at the surface would lead to an entire change of form, and therefore the constitution here supposed could not be permanent.

444. In the system of Saturn, we have a great deviation from the general order, which, nevertheless, has led to a very unexpected verification of some of the conclusions deduced above. A principle extremely like that which is the basis of all the foregoing reasonings, led one of the greatest philosophers of the present age to discover the revolution of Saturn's ring on its axis, and even to determine the velocity of that revolution, such as it has been since found by obšervation. Laplace, laying it down as a maxim, that nothing in nature can exist, where there are causes of change, not balanced or compensated by other causes, * concluded, that the parts of the ring must be held from falling down to the body of the planet by some other force than their mere cohesion to one another. Were it otherwise, every particle detached from the ring, by any means, must descend in a straight line, almost perpendicular to the surface of Saturn ; and the final destruction of the ring must be in-

---

* Laplace, *ubi supra*, p. 242.

evitable. The only force that could balance this effect of gravitation, seemed to be a centrifugal force, arising from the rotation of the ring on an axis passing through its centre, and perpendicular to its plane. Laplace proceeded to inquire what celerity of rotation was adequate to this effect, and found that one of ten hours and a quarter would be required, which is almost precisely the time afterwards determined by Dr Herschel from actual observation. If, with this rotation, the ring is a solid annulus generated by the rotation of a very flat ellipsis about a given point in its greater axis, coinciding with the centre of Saturn, it may be so constituted, that the attraction of Saturn, combined with the centrifugal force, may produce a force perpendicular to its surface, and may enable detached parts to remain at rest, animals, for instance, to walk on its surface, and fluids to be *in equilibrio*. The system of Saturn is thus fortified against the lapse of time, as effectually as that of the earth itself; and the means by which this is accomplished, seem to prove, that the weapons which time employs, are in both cases the same, viz. the slow wearing and decomposition of the solid parts. This slow wearing may have produced the figure by which its action is most effectually resisted.

445. Thus Dr Hutton's theory of the earth comes at last to connect itself with the researches of physical astronomy. The conclusion to be

drawn from this coincidence is to the credit of both sciences. When two travellers, who set out from points so distant as the mineralogist and the astronomer, and who follow routes so different, meet at the end of their journey, and agree in their report of the countries through which they have passed, it affords no slight presumption, that they have kept the right way, and that they relate what they have actually seen.

NOTE XXVI. § 133.

Prejudices relating to the Theory of the Earth.

446. Among the prejudices which a new theory of the earth has to overcome, is an opinion, held, or affected to be held, by many, that geological science is not yet ripe for such elevated and difficult speculations. They would, therefore, get rid of these speculations, *by moving the previous question,* and declaring that at present we ought to have no theory at all. We are not yet, they allege, sufficiently acquainted with the phenomena of geology; the subject is so various and extensive, that our knowledge of it must for a long time, perhaps for ever, remain extremely imperfect. And hence it is, that the theories hitherto proposed have succeeded one another with so great rapidity, hardly any

of them having been able to last longer than the
discovery of a new fact, or a fact unknown when it
was invented. It has proved insufficient to connect this fact with the phenomena already known,
and has therefore been justly abandoned. In this
manner, they say, have passed away the theories of
Woodward, Burnet, Whiston, and even of Buffon;
and so will pass, in their turn, those of Hutton and
Werner.

447. This unfavourable view of geology, ought
not, however, to be received without examination;
in science, presumption is less hurtful than despair,
and inactivity is more dangerous than error.

One reason of the rapid succession of geological
theories, is the mistake that has been made as to
their object, and the folly of attempting to explain
by them the first origin of things. This mistake
has led to fanciful speculations that had nothing
but their novelty to recommend them, and which,
when that charm had ceased, were rejected as mere
suppositions, incapable of proof. But if it is once
settled, that a theory of the earth ought to have no
other aim but to discover the laws that regulate the
changes on the surface, or in the interior of the
globe, the subject is brought within the sphere
either of observation or analogy; and there is no
reason to suppose, that man, who has numbered
the stars, and measured their forces, shall ultimately prove unequal to this investigation.

448. Again, theories that have a rational object, though they be false or imperfect in their principles, are for the most part approximations to the truth, suited to the information at the time when they were proposed. They are steps, therefore, in the advancement of knowledge, and are terms of a series that must end when the real laws of nature are discovered. It is, on this account, rash to conclude, that in the revolutions of science, what has happened must continue to happen, and because systems have changed rapidly in time past, that they must necessarily do so in time to come.

He who would have reasoned so, and who had seen the ancient physical systems, at first all rivals to one another, and then swallowed up by the Aristotelian; the Aristotelian physics giving way to those of Descartes; and the physics of Descartes to those of Newton; would have predicted that these last were also, in their turn, to give place to the philosophy of some later period. This is, however, a conclusion that hardly any one will now be bold enough to maintain, after a hundred years of the most scrupulous examination have done nothing but add to the evidence of the NEW-TONIAN SYSTEM. It seems certain, therefore, that the rise and fall of theories in times past, does not argue, that the same will happen in the time that is to come.

449. The multifarious and extremely diversified object of geological researches, does, no doubt, render the first steps difficult, and may very well account for the instability hitherto observed in such theories; but the very same thing gives reason for expecting a very high degree of certainty to be ultimately attained in these inquiries.

Where the phenomena are few and simple, there may be several different theories that will explain them in a manner equally satisfactory; and in such cases, the true and the false hypotheses are not easily distinguished from one another. When, on the other hand, the phenomena are greatly varied, the probability is, that among them, some of those *instantiæ crucis* will be found, that exclude every hypothesis but one, and reduce the explanation given to the highest degree of certainty. It was thus, when the phenomena of the heavens were but imperfectly known, and were confined to a few general and simple facts, that the Philolaic could claim no preference to the Ptolemaic system : The former seemed a possible hypothesis; but as it performed nothing that the other did not perform, and was inconsistent with some of our most natural prejudices, it had but few adherents. The invention of the telescope, and the use of more accurate instruments, by multiplying and diversifying the facts, established its credit; and when not only the general laws, but also the inequalities, and disturbances of the planetary motions were understood, all physical hypotheses vanished, like phantoms, before the philosophy of Newton. Hence the number, the variety, and even the complication of facts, contribute ultimately to separate truth from falsehood ; and the same causes which, in any case, render the first attempts toward a theory difficult, make the final success of such attempts just so much the more probable.

This maxim, however, though a general encouragement to the prosecution of geological inquiries, does not amount to a proof that we are yet arrived at the period when those inquiries may safely assume the form of a theory. But that we are arrived at such a period, appears clear from other circumstances.

450. It cannot be denied, that a great multitude of facts, respecting the mineral kingdom, are now known with considerable precision ; and that the many diligent and skilful observers, who have arisen in the course of the last thirty years, have produced a great change in the state of geological knowledge. It is unnecessary to enumerate them all ; Ferber, Bergman, Deluc, Saussure, Dolomieu, are those on whom Dr Hutton chiefly relied ; and it is on their observations and his own that his system is founded. If it be said, that only a small part of the earth's surface has yet been sur-

veyed, and described with such accuracy as is
found in the writers just named, it may be answered, that the earth is constructed with such a
degree of uniformity, that a tract of no very large
extent may afford instances of all the leading facts
that we can ever observe in the mineral kingdom.
The variety of geological appearances which a
traveller meets with, is not at all in proportion to
the extent of country he traverses ; and if he take
in a portion of land sufficient to include primitive and secondary strata, together with mountains, rivers, and plains, and unstratified bodies in
veins and in masses, though it be not a very large
part of the earth's surface, he may find examples
of all the most important facts in the history of
fossils.   Though the labours of mineralogists have
embraced but a small part of the globe, they may
therefore have comprehended a very large proportion of the phenomena which it exhibits ; and
hence a presumption arises, that the outlines, at
least, of geology have now been traced with tolerable truth, and are not susceptible of great variation.

451. When the phenomena of any class are in
general ambiguous, and admit of being explained
by different or even opposite theories ; if few of
those exclusive facts are known, which admit but
of one or a few solutions, then we have no right to
expect much from our endeavours to generalize,

except the knowledge of the points where our information is most deficient, and to which our observations ought chiefly to be directed. But that many of the exclusive and unambiguous instances are known, in the natural history of the globe, I think is evident from the reasoning in the foregoing pages, where so many examples have occurred of appearances that give the most direct negative to the Neptunian system, and exclude it from the number of possible hypotheses, by which the phenomena of geology can be explained. The abundance of such instances is an infallible sign, that the mass of knowledge is in that state of fermentation from which the true theory may be expected to emerge.

452. Another indication of the same kind, is the near approach that even the most opposite theories make, in some respects, to one another. There are so many points of contact between them, that they appear to approximate to an ultimate state, in which, however unwillingly, they must at last coincide. That ultimate form, too, which all these theories have a tendency to put on, if I am not deceived, is no other than that of the Huttonian theory.

453. The first example I shall take from the system of Saussure. It is to be regretted, that this excellent geologist has no where given us a complete account of his theory. Some of the leading principles of it are, however, unfolded in the course

of his observations, and enable us to form a notion
of its general outline. It was evidently far removed from the system of subterraneous heat, and
seems, especially in the latter part of the author's
life, to have been very much accommodated to the
prevailing system of Werner. Nevertheless, with
so little affinity between their general views, Saussure and Hutton agree in that most important article which regards the elevation of the strata.
Saussure plainly perceived the impossibility of the
strata being formed in the vertical situations which
so many of them now occupy ; and he takes great
pains to demonstrate this impossibility, from some
facts that have been referred to above. He also
believed that this elevation had been given to strata
that were originally level, by a force directed upwards, or by the *refoulement* of the beds, not by
their falling in, as is the opinion of Deluc and some
other of the Neptunists.

Now, whoever admits this principle, and reasons
on it consistently, without being afraid to follow it
through all its consequences, must unavoidably come
very close to the Huttonian theory. He must see,
that a power which, acting from below, produced
this great effect, can never have belonged to water,
unless rarefied into steam by the application of heat.
But if it be once admitted that heat resides in the
mineral regions, the great objection to Dr Hutton's
system is removed ; and the theorist, who was fur-

nished with so active and so powerful an agent, would be very unskilful in the management of his own resources, if he did not employ it in the work of consolidating as well as in that of raising up the strata.   A little attention will show, that it is qualified for both purposes; though insuperable objections must, no doubt, offer themselves, where the effects of compression are not understood.   We may safely conclude, then, that the accurate and ingenious Oreologist of Geneva ought to have been a *Plutonist*, in order to give consistency to the principles which he had adopted, and to make them coalesce as parts of one and the same system.   If he embraced an opposite opinion, it probably was from feeling the force of those objections that arise from our discovering nothing in the bowels of the earth like the remains left by combustion, or inflammation, at its surface.   The secret by which these seeming contradictions are to be reconciled, was unknown to this mineralogist, and he has accordingly decided strongly against the action of fire, even in the case of those unstratified substances that have the greatest affinity to volcanic lava.

454.   The theoretical conclusions of another accurate and skilful observer, Dolomieu, furnish a still more remarkable example of a tendency to union between systems professedly hostile to one another.

This ingenious mineralogist, observing the in-

terposition of the basalt between stratified rocks, so
that it had not only regular beds of sandstone for
its base, but was also covered with beds of the same
kind, saw plainly that these appearances were inconsistent with the supposition of common volcanic
explosions at the surface. He therefore conceived,
that the volcanic eruption had happened at the bottom of the sea, (the level of which, in former ages,
had been much higher than at present,) and that
the materials afterwards deposited on the lava, had
been in length of time consolidated into beds of
stone. It is evident, that this notion of submarine
volcanoes, comes very near, in many respects, to Dr
Hutton's explanation of the same appearances. If
the only thing to be accounted for were the phenomenon in question, it cannot be denied that Dolomieu's hypothesis would be perfectly sufficient ;
but Dr Hutton, to whom this phenomenon was
familiar, and who, like Dolomieu, conceived the
basalt to have been in fusion, was convinced that
the retreat of the sea was not a fact well attested by
geological appearances, and if admitted, was inadequate to account for the facts usually explained by
it. He conceived, therefore, that such lava as the
preceding had flowed not only at the bottom of the
sea, but in the bowels of the earth, and having
been forced up through the fissures of rocks already formed, had heaved up some of these rocks,
and interposed itself between them. This agrees

with the other facts in the natural history both of
the basaltes and the strata.

It is plain, that, in this, there is a great approach of the two theories to one another : both
maintain the igneous origin of basaltes, and its
affinity to lava ; both acknowledge that this lava
cannot have flowed at the surface, and that the
strata which cover it have been formed at the bottom of the sea. They only differ as to the mode
in which the submarine or subterraneous volcano
produced its effect, and that difference arises
merely from the one geologist having generalized
more than the other. Dolomieu sought to connect the basalt with the lavas that proceed from
volcanic explosions at the surface ; Dr Hutton
sought not only to connect these two appearances
with one another, but also with the other phenomena of mineralogy, particularly with the veins of
basaltes, and the elevation of the strata.

455. In another point, the coincidence of Dolomieu's opinions and Dr Hutton's is still more
striking. The former has remarked, that many
of the extinguished volcanoes are in granite countries, and that, nevertheless, the lavas that they
have erupted contain no granitic stones. There
must be, therefore, says he, something under the
granite, and this last is not, at least in all cases, to
be considered as the basis of the mineral kingdom,
or as the body on which all others rest. In this

system, therefore, granite is not always a primordial rock, any more than in Dr Hutton's.

But Dolomieu makes a still nearer advance to
the Huttonian theory ; for he supposes, that under the solid and hard crust of the globe, there is
a sphere of melted stone, from which this basaltic
lava was thrown up.   The system of subterraneous heat is here adopted in its utmost extent, and
in that form which is considered as the most liable
to objection, viz. the existence of it at the present
moment, in such a degree as to melt rocks, and
keep them in a state of fusion.   In this conclusion, the two theories agree perfectly ; and if they
do so, it is only because the nature of things has
forced them into union, notwithstanding the dissimilitude of their fundamental principles.

This ought to be considered as a strong proof,
that the phenomena known to mineralogists are
sufficient to justify the attempts to form a theory of
the earth, and are such as lead to the same conclusions, where there was not only no previous con-
cert, but even a very marked opposition.   I have
already observed, that there is a greater tendency
to agree among geological theories, than among the
authors of those theories.

456. Another circumstance worthy of consideration is, that in the search which the Neptunists
have made, for facts most favourable to the aqueous formation of minerals, we find hardly any of

8

a kind that was unknown to the author of the system here explained. The appearances on which Werner grounds his opinion with respect to basaltes, and by which he would exclude the action of fire from any share in the formation of it, are all comprehended in the alternation of that rock with beds, or strata obviously of aqueous origin. Now these appearances were well known to Dr Hutton, and are easily explained by his theory, provided the effects of compression are admitted. From this, and the other circumstances just observed, I am disposed to think, that the great facts on which every geological system must depend, are now known, and that it is not too bold an anticipation to say, that a theory of the earth, which explains all the phenomena with which we are at present acquainted, will be found to explain all those that remain to be discovered.

457. The time indeed was, and we are not yet far removed from it, when one of the most important principles involved in Dr Hutton's theory was not only unknown, but could not be discovered. This was before the causticity produced in limestone by exposure to fire was understood, and when it was not known that it arose from the expulsion of a certain aërial fluid, which before was a component part of the stone. It could not then be perceived, that this aërial part might be retained by pressure, even in spite of the action of fire, and

that in a region where great compression existed, the absence of causticity was no proof that great heat had not been applied. The discoveries of Dr Black, therefore, mark an era, before which men were not qualified to judge of the nature of the powers that had acted in the consolidation of mineral substances. Those discoveries were, indeed, destined to produce a memorable change in chemistry, and in all the branches of knowledge allied to it ; and have been the foundation of that brilliant progress, by which a collection of practical rules, and of insulated facts, has in a few years risen to the rank of a very perfect science. But even before they had explained the nature of carbonic gas, and its affinity to calcareous earth, I am not sure but that Dr Hutton's theory was, at least, partly formed, though it must certainly have remained, even in his own opinion, exposed to great difficulties. His active and penetrating genius soon perceived, in the experiments of his friend, the solution of those difficulties, and formed that happy combination of principles, which has enabled him to explain the most enigmatical appearances in the natural history of the earth.

As we are not yet far removed from the time when our chemical knowledge was too imperfect to admit of a satisfactory explanation of the phenomena of mineralogy, so it is not unlikely that we are approaching to other discoveries that are to throw

4

new light on this science.   It would, however, be
to argue strangely to say, that we must wait till
those discoveries are made before we begin any theoretical reasonings.   If this rule were followed, we
should not know where the imperfections of our
science lay, nor when the remedies were found out,
should we be in a condition to avail ourselves of
them.   Such conduct would not be caution, but ti-
midity, and an excess of prudence fatal to all philosophical inquiry.

458. The truth, indeed, is, that in physical inquiries, the work of theory and observation must go
hand in hand, and ought to be carried on at the same
time, more especially if the matter is very complicated, for there the clue of theory is necessary to direct
the observer. Though a man may begin to observe
without any hypothesis, he cannot continue long
without seeing some general conclusion arise ; and
to this nascent theory it is his business to attend,
because, by seeking either to verify or to disprove it,
he is led to new experiments, or new observations.
He is led also to the very experiments and observations that are of the greatest importance, namely,
to those *instantiæ crucis*, which are the *criteria* that
naturally present themselves for the trial of every
hypothesis.   He is conducted to the places where
the transitions of nature are most perceptible, and
where the absence of former, or the presence of
new circumstances, excludes the action of imagin-

ary causes. By this correction of his first opinion,
a new approximation is made to the truth ; and by
the repetition of the same process, certainty is finally
obtained. Thus theory and observation mutually
assist one another ; and the spirit of system, against
which there are so many and such just complaints,
appears, nevertheless, as the animating principle of
inductive investigation. The business of sound
philosophy is not to extinguish this spirit, but to restrain and direct its efforts.

459. It is therefore hurtful to the progress of
physical science to represent observation and theory
as standing opposed to one another. Bergman has
said, " Observationes veras quàm ingeniosissimas
fictiones sequi præstat ; naturæ mysteria potius in-
dagare quàm divinare."

If it is meant by this merely to say, that it is better to have facts without theory, than theory without facts, and that it is wiser to inquire into the se-
crets of nature, than to guess at them, the truth of
the maxim will hardly be controverted. But if we
are to understand by it, as some may perhaps have
done, that all theory is mere fiction, and that the
only alternative a philosopher has, is to devote himself to the study of facts unconnected by theory, or
of theory unsupported by facts, the maxim is as far
from the truth, as I am convinced it is from the real
sense of Bergman. Such an opposition between
the business of the theorist and the observer, can

only occur when the speculations of the former are vague and indistinct, and cannot be so *embodied* as to become visible to the latter. But the philosopher who has ascended to his theory by a regular generalization of facts, and who descends from it again by drawing such palpable conclusions as may be compared with experience, furnishes the infallible means of distinguishing between *perfect science* and *ingenious fiction.* Of a geological theory that has stood this double test of the analytic and synthetic methods, Dr Hutton has furnished us with an excellent instance, in his explanation of granite. The appearances which he observed in that stone led him to conclude, that it had been melted, and injected while fluid, among the stratified rocks already formed. He then considered, that if this is true, veins of granite must often run from the larger masses of that stone, and penetrate the strata in various directions; and this must be visible at those places where these different kinds of rock come into contact with one another. This led him to search in Arran and Glentilt for the phenomena in question; the result, as we have seen, afforded to his theory the fullest confirmation, and to himself the high satisfaction which must ever accompany the success of candid and judicious inquiry.

460. It cannot, however, be denied, that the impartiality of an observer may often be affected by system; but this is a misfortune against which the

want of theory is not always a complete security. The partialities in favour of opinions are not more dangerous than the prejudices against them; for such is the spirit of system, and so naturally do all men's notions tend to reduce themselves into some regular form, that the very belief that there can be no theory, becomes a theory itself, and may have no inconsiderable sway over the mind of an observer. Besides, one man may have as much delight in pulling down, as another has in building up, and may choose to display his dexterity in the one occupation as well as in the other. The want of theory, then, does not secure the candour of an observer, and it may very much diminish his skill. The discipline that seems best calculated to promote both, is a thorough knowledge of the methods of inductive investigation; an acquaintance with the history of physical discovery; and the careful study of those sciences in which the rules of philosophising have been most successfully applied.

END OF VOLUME FIRST.

Printed by George Ramsay & Co.
Edinburgh, 1821.

Printed in the United States
By Bookmasters